Geology of the Hudson Valley

A Billion Years of History

Steven H. Schimmrich

ISBN: 9798655926752
Independently published

To all those who wander the Hudson Valley and strive
to learn more about this beautiful world surrounding us.

Table of Contents

Acknowledgments

"And I know it seems easy," said Piglet to himself, "but
it isn't every one who could do it."

A. A. Milne, *The House at Pooh Corner*

Writing a book is a solitary endeavor yet could not be done without the assistance and inspiration of many people. I've been fortunate to have been influenced in my geological education by several outstanding teachers – Professor Lawrence "Bob" Matson of Ulster County Community College who first turned me on to geology, Dr. Russell Waines of SUNY New Paltz who taught me how to look at rocks in the field, Dr. Win Means of SUNY Albany who taught me that the simplest questions are often the most profound, and Dr. Stephen Marshak of the University of Illinois who introduced me to the Hudson Valley fold-thrust belt.

I am also grateful to the administration and trustees of Ulster County Community College for granting me a sabbatical leave to do much of the writing and research necessary to complete this book. I would also like to thank all of my students at SUNY Ulster who've taught me as much as I've hopefully taught them.

This book was greatly improved by a number of kind people who proofread sections for me (some numerous times). All of them contributed useful comments and criticisms for which I'm grateful.

I would like to thank Jennifer Wulfe for first encouraging me to write and putting up with me when I pursued that self-absorbed avocation. She is also an amazing artist who sketched many of the illustrations in this book and whose encouragement, comments, and criticisms helped turn a rough idea into a completed book.

Prelude
Ancient Seas

If by some fiat I had to restrict all this writing to one sentence, this is the one I would choose: The summit of Mt. Everest is marine limestone.

John McPhee, *Basin and Range*

I can take you to a rock outcrop on the side of the road in Kingston, a small city in the heart of New York's Hudson Valley. This outcrop looks no different from any of the other gray rocks on the side of the road, but if you pull over your car, walk up and down the outcrop, and take a closer look, you'll see something interesting. When I tell you to take a closer look, I mean exactly that. When I lead field trips of students to look at these rocks, they usually stand a good 20 feet away gazing at them as if possessed by some superhuman power to intuit their nature from afar. That doesn't work. You have to kick aside the weeds, ignore the ticks, and get up close to the rock. Touch it. Press your nose against the outcrop. Don't just get your hands dirty, get your face dirty. That's the only way to really look at rocks. What's there to see in this outcrop? If you take your time and look carefully you'll see a treasure trove of fossils. Fossils of marine shells, small ring-shaped fossils about the size of your fingernail, and, if you're persistent or just plain lucky, you may even find a fist-sized mass that resembles a petrified honeycomb.

The fossil shells are of marine creatures called brachiopods. Superficially, they look like modern bivalves – the clams, oysters, and scallops we're all familiar with from the seashore or our dinner plate –

but they're not closely related to them at all. Brachiopods actually belong to a completely different phylum (the highest level of classification within the animal kingdom). While these ancient animals lived in hinged shells of calcium carbonate like the bivalves, the creature inside was an entirely different sort of animal. If we could travel back in time to when these shells were filled with living animals, a little over 400 million years ago, we would see them gently floating in the ocean currents attached to the seafloor by a short, fleshy stalk called a pedicle. Inside the gaping shell, we would see a ciliated, horseshoe-shaped structure called a lophophore. Modern bivalves don't have a lophophore (if they did, they would be distinctly unpleasant to eat) but brachiopods used these organs to strain out small particles of organic material from the seawater – a process called suspension feeding.

Brachiopod shells also differ from modern bivalve shells in having a different symmetry. Brachiopod shells have a left-right mirror plane of symmetry while the top and bottom halves of the shell, called valves, differ in shape. Modern bivalves (think of a clam) usually have matching valves but generally lack this left-right symmetry.

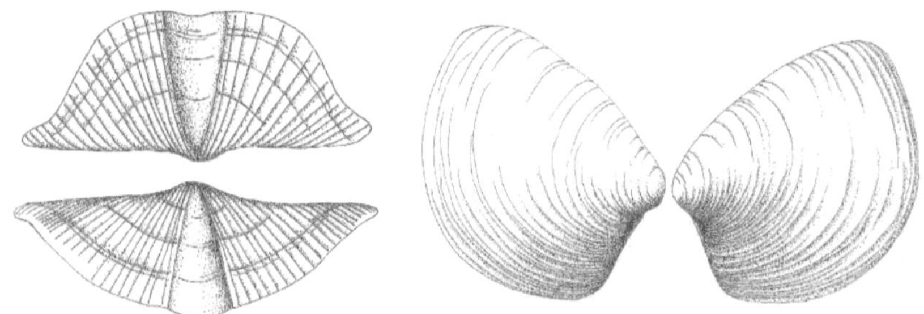

Symmetry of brachiopod (left) and bivalve (right) shells

If you look closely, carefully avoiding the ubiquitous poison ivy covering the outcrop, you'll see several thin layers of broken-up brachiopod shells. While there are a number of different species of brachiopods visible in this outcrop, one of the more abundant is named *Gypidula coeymanensis*. Like living animals, fossils are classified by the binomial system first proposed by the Swedish botanist Carl von Linné (1707-1778). Linné is better known by the Latinized version of his name – Carolus Linnaeus. In 1735, he published *Systema Naturae* in which he first proposed what later became the Linnaean classification system in which every organism is assigned to a unique two-word (binomial) species name. Taxonomists classify living organisms into a hierarchical system of kingdoms, phyla, classes, orders, families, genera, and species based upon the number of characteristics different organisms have in common. Two organisms belonging to the same genus, for example, share a lot more characteristics than two organisms only belonging to the same taxonomic class. If we examine the Linnaean classification of the *Gypidula coeymanensis* brachiopod, the woolly mammoth (another extinct organism), and modern humans, we see that we obviously share more physical characteristics with mammoths (a backbone, hair, mammary glands, etc.) than with a lowly brachiopod.

Kingdom	Animalia	Animalia	Animalia
Phylum	Brachiopoda	Chordata	Chordata
Class	Rhynchonellata	Mammalia	Mammalia
Order	Pentamerida	Proboscidea	Primates
Family	Gypidulidae	Elephantidae	Hominidae
Genus	Gypidula	Mammuthus	Homo
Species	*Gypidula coeymanensis*	*Mammuthus primigenius*	*Homo sapiens*

Linnaean classification of Gypidula, the woolly mammoth, and modern humans

One significant problem with classifying fossils, however, is that we're typically working with scant fragments of the once-living organism and the soft parts are usually entirely gone. Gone, like the *Gypidula*, for hundreds of millions of years. As a matter of fact, paleontologists have estimated that 99% of all brachiopod species are known to us only as fossils. Despite these difficulties, somewhere around 30,000 species of brachiopods have been formally described while fewer than 400 exist today.

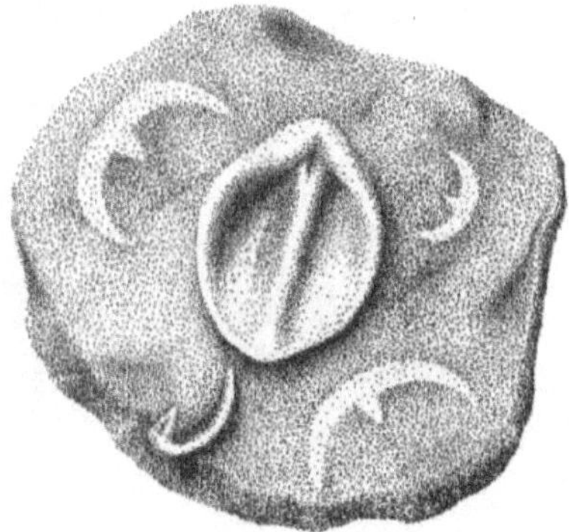

Gypidula coeymanensis shells in face view and cross section

The *Gypidula* shells in our roadside rock outcrop are easy to recognize because they often appear in cross-section as crescent moon shapes with small spikes protruding from the center of the crescent. *Gypidula* also had thick shells which appear to be adapted for life in turbulent, shallow seawater. The turbulent nature of their ancient environment is also attested to by the numerous broken shell fragments visible in these rock layers.

The small ring-shaped fossils scattered throughout the outcrop are the stem fragments, called columnals, of animals known as crinoids. Crinoids (it's hard to tell the exact species from a few columnal fragments) are sometimes called sea lilies due to their resemblance to flowers, but they're actually echinoderms and related to animals like starfish and sea urchins. Imagine an upside-down starfish on top of a stalk attached to the seafloor. Fold up the legs of the starfish into a bulb and imagine feathery appendages sprouting out of the top and waving in the ocean currents. Crinoids, like brachiopods, are suspension feeders ingesting organic material floating in the seawater with the assistance of these feathery arms. The bulbous head of a crinoid, called the calyx, is composed of numerous tiny plates of calcium carbonate held together by soft tissue. After the crinoid dies, the soft tissue rots away and the tiny plates are scattered far and wide by the bottom currents. The more sturdy columnals are often the only thing preserved as fossils – especially in the turbulent, shallow waters where we now know these animals lived (which we determined by looking at the robust *Gypidula* shells).

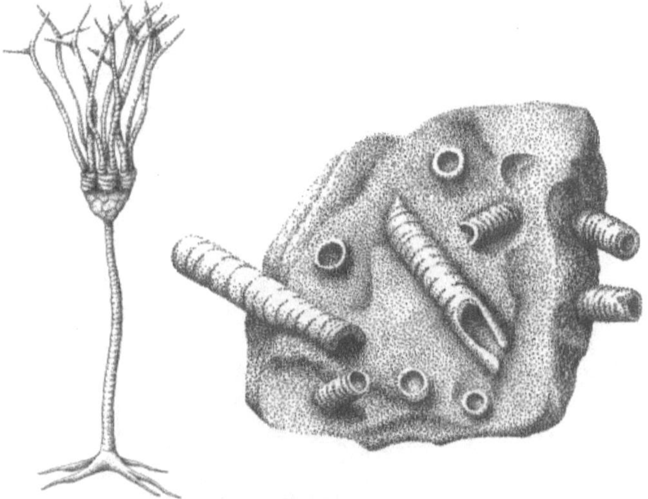

Crinoid as it was in life and as columnal fossils

If you're lucky enough to find a petrified 'honeycomb' in the outcrop, you're actually looking at an ancient coral called *Favosites helderbergensis* – otherwise known, not surprisingly, as honeycomb coral. Within each of the little holes in this mass of coral lived tiny animals called polyps with extended tentacles waving in the ocean currents. These tentacles differed from the arms of the crinoids in that they had stinging cells which enabled them to stun and capture their next meal (typically plankton). The specialized stinging cells on the tentacles of corals are called cnidocytes and are also found on jellyfish and sea anemones which all belong to phylum cnidaria (the C is silent). Cnidarians are among the earliest multicellular organisms existing on Earth – some of the oldest animal fossils are jellyfish impressions over 600 million years old from southern Australia. *Favosites* is much younger than these earliest cnidarians, a mere 400 million years old, and this colonial coral was an important builder of small, patchy reefs in these ancient seas.

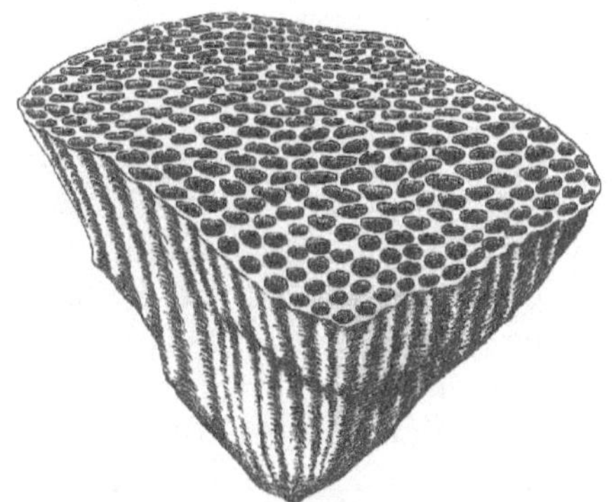

Favosites helderbergensis – also known as honeycomb coral

The hard parts of the honeycombed *Favosites* coral, like the shells of the brachiopods and the columnals of the crinoids, were also composed of calcium carbonate. Many marine invertebrates, even today as they were 400 million years ago, have hard parts composed of calcium carbonate ($CaCO_3$), a mineral known as aragonite or calcite depending upon its crystalline structure. What happens when all of these animals die? Their hard parts eventually disintegrate into a fine-grained sediment known as carbonate mud. What happens when carbonate mud lithifies into rock? It forms beds of the sedimentary rock limestone. These fossils you're looking at are all entombed in beds of limestone – the carbonate mud of an ancient seafloor. The rock itself is a fossil – the remains of ancient life.

These fossils represent once-living marine invertebrates that thrived in a shallow, subtropical sea. While fossil corals and other marine invertebrates from this time were in some ways structurally different from modern organisms, it's reasonable to assume that they also preferred sunlit, warm, relatively-clear ocean waters to thrive. Swim among coral reefs in the Florida Keys and you'll have a rough idea of what the mid-Hudson Valley looked like almost half a billion years ago. How did these animals come to be fossilized here? Remember that we're in the heart of the Hudson Valley. Even if we flooded this area with seawater, we're too far from the equator and the water would be too cold to support this type of marine life. Reread the quotation from John McPhee at the beginning of this chapter. The fossils we find in Kingston, and elsewhere in the Hudson Valley, are just as surprising and out of place as the marine limestones at the summit of Mount Everest. While it's highly unlikely that anyone reading this will ever see the marine limestones topping Mount Everest (in person, at least), we can all visit the marine limestones in the heart of the Hudson Valley. All it takes is for you to pull your car over to the side of the road and be willing to get your face a little dirty.

This book was written to provide an overview of the long geologic history of the beautiful Hudson Valley region of New York; an area I call home. As with all broad surveys, many of the technical details are simplified or omitted to make this information accessible to everyone. I would also ask the reader to keep in mind that the details of any single rock formation can be quite complex and geologists still argue over many of them at the outcrop (or later over pints of beer in the local pub).

The best way to learn about the geology of the Hudson Valley would be for you to travel with me on a series of field trips, as do my students, but it's obviously not possible for everyone to have this pleasure. This book is therefore a second-best option for learning our local geology since, as any geologist will tell you, the only real way to learn is out on the rocks. My hope is that as we tour the billion years or so of geologic history in the Hudson Valley within these pages, you'll gain a new appreciation for those previously nondescript rock outcrops at the side of the road that you drive past in your car every day. Maybe you'll even visit some of these outcrops, give them a closer look, and accord them the respect they deserve as ancient ambassadors to our modern world.

1
In The Beginning

To be ignorant of what occurred before you were born
is to remain always a child.

Marcus Tullius Cicero, *Orator*

As a geologist, it always surprises me when I talk with people who believe that the entire universe arose during six days of divine creation a few thousand years ago. Some refer to the age calculated by Anglican Archbishop James Ussher (1581-1656) of 4,004 BCE* for the *creatio ex nihilo* event described in the book of Genesis. Older King James Version Bibles even place this 4,004 BCE date as a footnote to Genesis 1:1 – "In the beginning God created the heaven and the earth." These young-Earth creationists also believe that much of the sedimentary rock and fossil record formed as a result of the global flood described a bit later in the book of Genesis. These beliefs, of course, are firmly rooted in a literalist interpretation of the Christian Bible and are completely rejected by modern science (as well as many, if not most, religious denominations). As a matter of fact, these ideas have been rejected by geologists for over 150 years now!

It's a failure of our educational system that so many people today have no concept of the age of the Earth or even the basic tenets of

---

* *The acronyms* BCE *(Before Common Era) and* CE *for (Common Era) are increasingly used instead of* BC *(Before Christ) and* AD *(anno Domini – the year of our Lord) in recognition that not everyone shares the Christian belief system.*

modern science. While most Americans are not young-Earth creationists, studies do consistently show a low level of scientific literacy among the general public. To a geologist, saying that the Earth is only a few thousand years old sounds as ludicrous as saying that the Earth is flat or that the Sun and all of the other planets actually orbit the Earth – ideas once supported, incidentally, by a literal interpretation of Scripture. One of the more prominent young-Earth creationist organizations recommends that schoolchildren challenge their teachers with the statement "Were you there?" when claims are made of an ancient Earth or the evolution of life on Earth through time. In other words, how can science say something about the ancient world when they also claim that no humans were around to record these events?

"Were you there," is a specious challenge – one might as well argue the veracity of any well-accepted historical fact including those dear to the heart of Scriptural literalists. A far more reasonable question to ask scientists, in my opinion, is "How do you know?" As a college professor, I wish more of my students would ask that question from time to time instead of simply copying down everything I say as if it were knowledge handed down from on high.

Unfortunately, in a book such as this, as in an undergraduate science class, time constraints prevent one from rigorously demonstrating the basis of all we believe to be "true" in science. Much of what I'll present in these pages must be accepted on good faith unless one is willing to spend years studying the science behind such claims. I hope, however, that the attentive reader will also begin to understand how geologists go about studying the Earth and how claims like "The Hudson Valley was covered by a shallow subtropical sea 400 million years ago" is completely supportable by physical evidence that anyone can go out and observe.

Like all good stories, we'll start at the beginning. Modern astronomy tells us that time and space itself had a beginning in the Big Bang almost 14 billion years ago. I'll leave the details of that event for my astronomer colleagues to explain but, suffice it to say, there is abundant and compelling observational evidence for this dramatic event. For nearly 10 billion years, the universe existed just fine without the Earth as our solar system had yet to form. A little less than 5 billion years ago, a cloud of gas and dust, called the solar nebula, collapsed in on itself (perhaps pushed by shock waves from a nearby supernova) and the hydrogen gas in the core of this contracting nebula eventually reached a high enough density to initiate thermonuclear fusion – the process by which stars produce energy by the fusion of hydrogen, the most abundant element in the universe, into helium, the second-most abundant element. Our sun was born as a young star in an already ancient universe.

The remaining disk of gas and dust orbiting our young sun eventually condensed into the various planets, moons, asteroids, and comets which form our present-day solar system. This solar nebula theory has some important observational support; astronomers have located protoplanetary disks of gas and dust around young stars in the Orion Nebula which are interpreted to be young solar systems in the process of planetary formation. Thousands of extrasolar planets have also been detected orbiting nearby stars within our galactic neighborhood with new ones discovered on an almost daily basis. Solar systems appear to be relatively common within our Milky Way galaxy.

By 4.6 billion years ago, our solar system existed since we have found meteorites on the Earth which date close to that age (these meteorites formed elsewhere in the solar system and fell to Earth much more recently). Soon after its formation, the early Earth began to replace its primordial atmosphere of hydrogen and helium, light elements which quickly leaked off into space, with an atmosphere of volcanic gases emitted from the interior of the planet as it cooled.

This earliest eon of geologic time is called the Hadean by geologists. The name comes from the Greek word *Hades* or Hell. This is an appropriate name since the early Earth was indeed a hellish environment. Astronomers believe our moon formed during this time by the slightly off-center collision of a large Mars-sized object with the early Earth. Astronomers have named this rogue planetoid Theia since, in Greek mythology, she gave birth to the moon goddess Selene. The massive collision would have completely melted the Earth and ejected a large volume of material into orbit which rather quickly (perhaps in only a few years) coalesced into our moon. During the Hadean Eon, the Moon was much closer to the Earth and would have looked 20 times larger in the sky. This is about the size of your fist held at arm's length – go out some night and compare that to the size of the modern-day full Moon which is smaller than the fingernail on your little finger at arm's length.

The Earth was also rotating faster then and a complete day was only about 6 hours long. Tides in the earliest oceans would have been impressive, over a kilometer high and washing completely over the low-lying, barren, proto-continents of the time. The Moon is gradually moving away from the Earth – laser reflectors on the Moon left by the Apollo astronauts reveal a recessional rate of 3.8 centimeters (1.5 inches) per year. If you're fortunate enough to live to be 80 years old, the Moon will be about 10 feet further away from Earth than it was on the day you were born. This recession of the Moon is currently slowing the rotation of the Earth by about 2 milliseconds per century through a process physicists call the conservation of angular momentum. This doesn't sound like much, but it adds up over the long eons of geologic time and corresponds to a loss of about one minute every three million years.

It's interesting to speculate what the Earth would have been like had the Moon never formed by this chance collision early in the history of the solar system. The rotational rate of the Earth would have decreased much more slowly without the braking effect of the Moon and days would still be shorter than 10 hours long. This increased rate of rotation would lead to much more powerful winds whipping across the surface making it extremely difficult for terrestrial life to evolve. There would be small tides from the gravitational effects of the Sun but the mega-tides of the Hadean Eon would not have occurred. Without larger tides, the physical and chemical properties of the oceans may have been very different from what they are today. If life had developed on such an Earth, it would likely have taken a very different path. This is one of the lessons to take away when studying Earth history – our present-day world is, in many respects, the end result of a number of chance events.

Fragments from the collision which formed the Moon, as well as debris left over from the formation of our solar system, led to numerous meteorite impacts on the young Earth. Some of these impacts may have been large enough to completely melt the Earth's earliest crust. The craters we see on the Moon today are remnants of this early bombardment. On Earth, craters from this time are long gone – erased by our planet's dynamic crustal movements and erosional processes.

The interior of the Hadean Earth was hotter than it is today. Part of that heat was from the kinetic energy imparted by collisions – the very collisions that assembled the planet from smaller dust and rock fragments orbiting within the solar nebula. Another portion of the heat was due to the radioactive decay of short-lived isotopes scattered within the interior of the planet. As radioactive isotopes decay over time, they release heat energy and their abundance in the interior of the Earth gradually decreases. This primordial heat led to the differentiation of the early Earth into a core, mantle, and crust. Iron, being especially abundant in our solar system and denser than most Earth materials, sank to the center to form the core of the Earth. The once liquid core has partially solidified today into a solid inner core and liquid outer core. Movement of liquid iron within the Earth's outer core generates the magnetic field presently surrounding the Earth. This is a fortunate occurrence since, without our magnetic field, the Earth would be bathed in a storm of charged particles from the Sun and this would have prevented a stable atmosphere from forming leading to a barren, rocky world. We see such a world today when we look at Mars which has a solidified iron core.

One consequence of a hotter Earth was more energetic volcanic activity. Volcanoes can be thought of as pressure valves in the Earth's crust since they act as a mechanism for transferring heat from the interior of the Earth to its surface. Volcanoes, however, were a good thing for the Hadean Earth. Even though it was a hellish place over 4 billion years ago, the intense volcanic activity contributed several important things to our evolving planet.

When volcanoes erupt they release large amounts of gas. The two dominant gases released by volcanoes are water vapor (H_2O) and carbon dioxide (CO_2). Minor gases include sulfur dioxide (SO_2), hydrogen sulfide (H_2S), nitrogen (N_2), and argon (Ar), among others. While the exact composition of the Earth's earliest atmosphere is the subject of some debate, these gases are thought to be major players. Note the lack of molecular oxygen (O_2) which comprises 21% of our atmosphere today. The Earth's earliest atmosphere had no oxygen and we can actually see evidence of this in ancient rocks.

Rocks from this earliest period of Earth history do not contain oxidized iron but instead contain iron compounds like iron sulfide which forms the mineral pyrite (FeS_2). Iron is a common element present in most sediments and will quickly oxidize to form the iron oxide mineral hematite (Fe_2O_3) when exposed to the Earth's modern oxygen-rich atmosphere. The beautiful "redbeds" of sandstone seen in the Canyonlands country of Arizona and Utah are red because of oxidized iron. Similarly, the red sandstones and shales seen here in the Catskill Mountains or Rockland County lowlands of the Hudson Valley are also colored by hematite. But the most ancient rocks on Earth, not found in New York State by the way, completely lack this mineral. In addition, the present-day atmospheres of both Venus and Mars, our next-door neighbors, have atmospheres of primarily carbon dioxide with lesser amounts of nitrogen, argon, and other volcanic gases. This is pretty much what we would expect from an atmosphere formed by volcanic outgassing and we analogously assume that this was what the Earth's early atmosphere was like as well.

It's actually a good thing that carbon dioxide was so abundant in the Earth's earliest atmosphere. Astronomers believe that when the solar system first formed, the Sun was 20-25% dimmer than it is today. The earliest Earth would have been a spinning ball of ice without the climate-modifying effects of CO_2. Carbon dioxide is well-known as a greenhouse gas acting to retain heat energy in the Earth's atmosphere. Some scientists speculate that surface temperatures on the Hadean Earth may have been several times what they are today due to this high level of CO_2 in the atmosphere. Not only has life evolved on Earth through time, our atmosphere has evolved as well.

What about the water vapor released by volcanic outgassing? On Venus, a small difference in the distance from the Sun between our sister planet and the Earth led to water vapor being broken apart into its constituent elements in the hot, thick atmosphere and disappearing forever. Planetary missions have shown, on the other hand, that water once flowed freely over the surface of Mars and is still present in the form of subsurface ice. Mars, however, also followed a different evolutionary path and now has too thin an atmosphere to support liquid water on the surface. On Earth, of course, water is also present but we're fortunate enough to have a range of atmospheric pressures and temperatures such that it can exist as vapor, liquid, or solid. The volcanic water vapor in the Earth's earliest atmosphere simply condensed to form liquid water oceans as soon as the crust cooled enough to support it. There may also have been a significant influx from the collisions of comets (often described as dirty snowballs due to their large percentage of water ice) early in the Earth's history. There are hints that liquid water oceans may have formed as early as 4.4 billion years ago according to the chemical analysis of minerals recovered from ancient rocks in Australia.

These earliest oceans were different from our modern-day ones in that they would have been initially fresh water and acidic rather than saline and alkaline as they are today. The acidity was from the extremely acid rain that would have formed in the Earth's earliest atmosphere since a carbon dioxide atmosphere will react with water vapor to form carbonic acid ($CO_2 + H_2O \rightarrow H_2CO_3$). In addition to making the oceans acidic, this acid rain would also have chemically weathered rocks at a faster rate than they are today. This increased chemical weathering resulted in plenty of free ions making their way into the ancient freshwater oceans gradually increasing their salinity. The Earth's seawater, like its atmosphere, has also evolved to its present state over geologic time.

One consequence of acidic oceans on the early Earth is the complete lack of limestone from this period of time. Limestone forms today in alkaline seawater but could not have formed in the acidic oceans of the early Earth. Rocks not only record evidence of ancient life, but also evidence of ancient environments on our planet.

We now leave the Hadean and move into the next eon of Earth history – the Archean (from the same Greek root as the word "archaic"). While microscopic crystals of a mineral called zircon have been found in Australia dating back 4.4 billion years, the surface of the Earth didn't calm down enough to preserve its earliest crustal rocks until a few hundred million years later. The oldest confirmed rock on Earth belongs to the Acasta Gneiss Complex and is found north of the town of

Yellowknife in northwestern Canada. This metamorphic rock has been radioactively dated to right around 4 billion years old (different parts of the complex have slightly differing ages). As I write this, however, a more recent discovery from ancient rocks in northern Quebec on the eastern shore of Hudson Bay may push the age of the oldest rock back to 4.28 billion years but this has yet to be confirmed.

Radioactive dating is a wonderful tool for geologists and the topic is worthy of a short digression. Atoms, as we all learned in elementary school, are composed of positively-charged protons and electrically-neutral neutrons residing in a central nucleus. Orbiting the nucleus are smaller negatively-charged particles called electrons. Keep in mind that "orbiting" doesn't mean that the electrons are little balls circling around the nucleus in the same way planets orbit a star but rather than the electrons are found somewhere within a series of three-dimensional shells around the nucleus.

While protons and neutrons are about the same size, electrons are 1,836 times smaller. The number of protons in the nucleus determines the kind of element the atom is: 1 proton makes hydrogen, 2 helium, 6 carbon, 26 iron, and 92 uranium, for example. This is truly amazing when you think about it – an atom with 16 protons is sulfur, an innocuous yellow powder, while an atom with 17 protons (one additional tiny particle) is chlorine, a poisonous, highly-reactive, green gas! The number of protons is the atomic number and determines where the element resides on the periodic table.

While the number of protons determines the element, the number of neutrons in an atom is somewhat flexible. Some atoms of the same element have differing numbers of neutrons, but the same number of protons, and are called isotopes (Greek for "equal place" since they occupy the same place on the periodic table). These alternate forms of the same element differ only in being a bit heavier or lighter based upon whether or not they have more or less neutrons in their nuclei and are called, not surprisingly, heavy or light isotopes. Isotopes are classified by their mass number – the number of protons (p^+) and neutrons (n^0) they possess in the nucleus. Since electrons (e^-) are so much smaller than protons and neutrons we can often ignore them when thinking about an atom's mass. The element carbon, for example, has three naturally-occurring isotopes with mass numbers of 12, 13, and 14.

Since carbon holds the number 6 position on the periodic table, this is its atomic number denoting the number of protons found in any carbon atom anywhere in the universe. Since there are three isotopes of carbon, they must therefore differ in the number of neutrons they possess. Carbon-12 is the standard form of carbon with 6 protons and 6 neutrons in its nucleus (almost 99% of all carbon is ^{12}C), ^{13}C is a heavy

Periodic Table of the Elements

Period	1	2	3	4	5	6	7	8	9	10	11	12	13	14	15	16	17	18
1	1 H Hydrogen																	2 He Helium
2	3 Li Lithium	4 Be Beryllium											5 B Boron	6 C Carbon	7 N Nitrogen	8 O Oxygen	9 F Fluorine	10 Ne Neon
3	11 Na Sodium	12 Mg Magnesium											13 Al Aluminum	14 Si Silicon	15 P Phosphorous	16 S Sulfur	17 Cl Chlorine	18 Ar Argon
4	19 K Potassium	20 Ca Calcium	21 Sc Scandium	22 Ti Titanium	23 V Vanadium	24 Cr Chromium	25 Mn Manganese	26 Fe Iron	27 Co Cobalt	28 Ni Nickel	29 Cu Copper	30 Zn Zinc	31 Ga Gallium	32 Ge Germanium	33 As Arsenic	34 Se Selenium	35 Br Bromine	36 Kr Krypton
5	37 Rb Rubidium	38 Sr Strontium	39 Y Yttrium	40 Zr Zirconium	41 Nb Niobium	42 Mo Molybdenum	43 Te Technetium	44 Ru Ruthenium	45 Rh Rhodium	46 Pd Palladium	47 Ag Silver	48 Cd Cadmium	49 In Indium	50 Sn Tim	51 Sb Antimony	52 Te Tellurium	53 I Iodine	54 Xe Xenon
6	55 Cs Caesium	56 Ba Barium	57-71 La-Lu Lanthanides	72 Hf Hafnium	73 Ta Tantalum	74 W Tungsten	75 Re Rhenium	76 Os Osmium	77 Ir Iridium	78 Pt Platinum	79 Au Gold	80 Hg Mercury	81 Tl Thallium	82 Pb Lead	83 Bi Bismuth	84 Po Polonium	85 At Astatine	86 Rn Radon
7	87 Fr Francium	88 Ra Radium	89-103 Ac-Lr Actinides	104 Rf Rutherfordium	105 Dd Dubnium	106 Sg Seaborgium	107 Bh Bohrium	108 Hs Hassium	109 Mt Meitnerium	110 Ds Darmstadtium	111 Rg Roentgenium	112 Cn Copernicium	113 Uut Ununtrium	114 Fl Flerovium	115 Uup Ununpentium	116 Lv Livermorium	117 Uus Ununseptium	118 Uuo Ununoctium

Lanthanide Series

57 La Lanthanum	58 Ce Cerium	59 Pr Praseodymium	60 Nd Niodymium	61 Pm Promethium	62 Sm Samarium	63 Eu Europium	64 Gd Gadolinium	65 Tb Terbium	66 Dy Dysprosium	67 Ho Holmium	68 Er Erbium	69 Tm Thulium	70 Yb Ytterbium	71 Lu Lutetium

Actinide Series

89 Ac Actinide	90 Th Thorium	91 Pa Protactinium	92 U Uranium	93 Np Neptunium	94 Pu Plutonium	95 Am Americium	96 Cm Curium	97 Bk Berkelium	98 Cf Californium	99 Es Einsteinium	100 Fm Fermium	101 Md Mendelevium	102 No Nobelium	103 Lr Lawrencium

Note: Elements in gray are not naturally occurring.

isotope of carbon containing one extra neutron, and ^{14}C is a heavier isotope of carbon containing two extra neutrons. Even though they're all carbon, the presence of one or more extra neutrons in the nucleus leads to surprisingly different properties.

Here's what's extraordinary about these three forms of carbon. The ^{12}C and ^{13}C forms of carbon are both called stable isotopes since they never change into any other elements (stable isotopes do have their uses in geology and we'll discuss these later when looking at climate change during the glacial ice age). Carbon-14, on the other hand, is unstable and will spontaneously emit subatomic particles causing it to change into a completely new element. This new element, nitrogen-14, has 7 protons (it's number 7 on the periodic table) and 7 neutrons. It's the standard, stable form of nitrogen. To change unstable ^{14}C into stable ^{14}N, a neutron within the ^{14}C nucleus emits an electron (along with a particle called an electron antineutrino) and changes into a proton within the nucleus of the new ^{14}N atom. This is called beta emission and is one of the processes by which atoms radioactively decay.

Isotopes like ^{14}C are called radioisotopes because of this property of radioactive decay. The truly amazing thing about radioactive decay is that even though it's a random process, meaning that we can't predict exactly when a particular isotope will decay, a large group of radioactive isotopes will break down at a constant measurable rate called the decay constant. When I talk about large groups of atoms, keep in mind that a cube of graphite (pure carbon) the size of the period at the end of this sentence (0.5 mm^3) can contain well over a billion billion (10^{18}) carbon atoms.

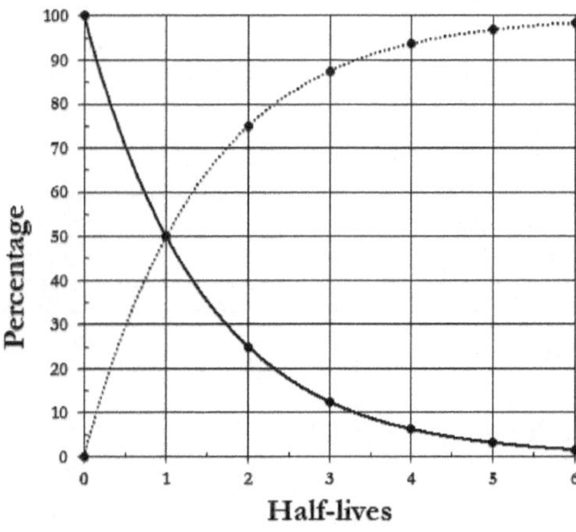

Exponential decay of radioactive ^{14}C (solid line) into ^{14}N (dotted line)

The decay of radioactive isotopes is typically expressed by a number related to the decay constant called the half-life which is the amount of time it takes for half of the radioisotopes to decay. The half-life of ^{14}C is 5,730 years. This means that if we start with 100% ^{14}C, we will have 50% ^{14}C and 50% ^{14}N after 5,730 years. After two half-lives, 11,460 years, we'll have 25% ^{14}C and 75% ^{14}N. When graphed, radioactive decay is represented by an exponential curve (solid line is decay of ^{14}C, dashed line is increase in ^{14}N).

Extensive research, conducted since the process of radioactivity was first discovered in 1896 by French scientist Antoine Henri Becquerel, has shown that the decay constant for radioisotopes is indeed constant. Given the tiny size of atomic nuclei compared to the outside environment, and the magnitude of the forces governing nuclear decay at these scales, the decay constant is completely unaffected by the physical state of the material (solid, liquid, or gas), the temperature, pressure, magnetic field, or any other known outside force. Because of this constant decay rate, radioisotopes can be used as timepieces to date various materials.

A limitation of radiometric dating is that after about 10 half-lives, there's less than 0.1% of the original radioisotope remaining and it becomes nearly impossible to obtain accurate dates. That's why ^{14}C is not used to date rocks, its half-life of 5,730 years means that it's essentially ineffective on objects older than about 60,000 years (and virtually all rocks are far, far older than this). As we'll discuss later, however, ^{14}C can be useful in geology since it can be applied when dating material from the end of the last ice age. Fortunately, there are a number of other radioisotopes with much longer half-lives which can be used to date rocks.

Parent Isotope	Daughter Isotope	Half-Life
Carbon-14	Nitrogen-14	5,730 yr
Uranium-237	Lead-207	7.04×10^8 yr
Potassium-40	Argon-40	1.25×10^9 yr
Uranium-238	Lead-206	4.47×10^9 yr
Rubidium-87	Strontium-87	48.8×10^9 yr
Samarium-147	Neodymium-143	10.6×10^{10} yr

Radioactive isotope pairs useful in dating geologic materials

Different radioisotopes are found in different minerals within different types of igneous and metamorphic rocks. The radiometric dating method used for a particular sample is governed by a number of factors and is beyond the scope of this book. It will have to suffice to say that after a century of study and technological refinement, geoscientists have a high degree of confidence in these methods.

Returning to our main story, the Acasta Gneiss is known to be about four billion years old because it contains small amounts of the radioisotope ^{238}U. Analysis of the uranium and lead isotopes within this rock allows geologists to constrain its age of formation. The Acasta Gneiss is a piece of continental crust and, geologically-speaking, the story of the Archean Eon is the story of the earliest evolution of the Earth's continental crust. Early in the Earth's history, two different types of crust developed – continental and oceanic. Oceanic crust underlies all of the world's oceans and is composed of basalt. Basalt is simply hardened lava, the same as you see pouring out of Hawaiian volcanoes today. It forms when the Earth's crust fractures and the mantle, the layer below the crust, partially melts and forms masses of magma which ascend and cool to form the basaltic sea floor. Basalts are dark gray to black, fairly dense, and contain abundant iron- and magnesium-rich (ferromagnesian) minerals and no quartz.

Continental crust, on the other hand, forms from a long cycle of crystallization, partial melting, and recrystallization of igneous rocks and has a composition more like granite. Granites are typically white or pinkish in color, less dense than basalts, and contain abundant grains of quartz and feldspar (quartzofeldspathic) minerals. Granitic crust, on average, is also about 5 times thicker than oceanic crust which, along with its lower density, allows it to "float" higher in the underlying mantle and results in it emerging above the surface of the global ocean today. While oceanic crust formed fairly quickly on the young Earth, continental crust grew over longer periods of time as small patches of granitic crust formed, collided and accreted to other fragments, and gradually grew into larger proto-continental blocks. They are called proto-continental blocks because they were significantly smaller than our modern-day continents. These earliest blocks of continental crust differed in some ways from the types of rocks forming today but none are exposed in New York so we won't worry any more about them.

Besides the formation of proto-continents, other interesting things were occurring on the Archean Earth. While there's indirect chemical evidence from ancient rocks exposed in Greenland that life was already in existence around 3.8 billion years ago, it's somewhat controversial and certainly not conclusive. By about 3.5 billion years ago, however, we do start to find some of the earliest true fossils from the Warrawoona Group of rocks exposed in western Australia. Note that a billion years have now passed since the Earth first formed. Let me repeat that – the Earth existed for a billion years before the earliest fossils appeared in the geologic record. There's a lot of time available for the origin of life on our young planet!

How did life develop on early Earth? The hypothesized process by which life first appeared is termed abiogenesis. Assuming that life arose through naturalistic processes and not through divine intervention, an axiomatic assumption in science which you may or may not like given your religious beliefs, life presumably developed through an earlier process of prebiotic chemical evolution. We know, for example, that some of the basic building blocks of life, such as amino acids, readily form given the right conditions. Amino acids are easily created in the laboratory under early Earth conditions and have even been detected in meteorites. Linking these basic building blocks together into more complex organic molecules like proteins, however, is a bit more difficult and problematic for researchers working on this question. A number of scenarios have been proposed, but all serious models involve life originally developing in the Earth's oceans. One compelling reason for this belief is that the lack of oxygen in the Earth's earliest atmosphere meant that there was no ozone layer (ozone is a form of molecular oxygen) and any land surfaces on Earth would have been bathed in sterilizing ultraviolet radiation from the Sun. A thin layer of water effectively blocks this deadly radiation from fragile early organisms.

However it happened, we know that the earliest undisputed life on Earth was microscopic and composed of single cells without a nucleus (as are modern bacteria). These types of organisms are called prokaryotes and they exclusively ruled the Earth for over 2 billion years (and are still pretty successful today as anyone who has weathered a bacterial infection knows). One ancient lineage of bacteria-like organisms is known as the Archaea and these organisms are often referred to as extremophiles due to their occurrence in harsh environments today. Extremophiles can be found in boiling hot springs (thermophiles), Antarctic ice (cryophiles), arid deserts (xerophiles), hypersaline lagoons (halophiles), acidic volcanic lakes (acidophiles), alkaline soda lakes (alkaliphiles), and even deep within the basaltic ocean crust (endoliths). NASA sterilizes their spacecraft because of fears that these hardy organisms could perhaps hitch a ride to places like Mars and survive (if not thrive) making our search for possible indigenous Martian life much more difficult.

Extremophiles are typically anaerobic organisms which do not require oxygen for life and utilize a metabolic process called chemosynthesis. Chemosynthetic organisms obtain their energy directly from chemicals existing in the natural environment. For example, some modern Archaea living near hydrothermal vents on the deep ocean seafloor obtain all of their energy directly from hydrogen sulfide gas released from these volcanic vents. This is what biologists believe the earliest life on Earth looked like – microscopic, single-celled bacteria thriving in the extreme environments of the Archean Earth. Fossils of these earliest organisms are rare and difficult to find due to their microscopic size and lack of hard parts, but they are known from a number of Archean Eon rock outcrops found at various locales around the world.

A little easier to find than the microscopic fossils of Archaea are odd objects the size and shape of cabbage heads found in some ancient sedimentary rock outcrops. These fossils, more properly called stromatolites, are actually laminated mounds formed from sticky mats of cyanobacteria trapping successive layers of seafloor sediment over time. They were important reef-builders in the ancient world long before the evolutionary development of corals. These cyanobacteria, which are sometimes erroneously called blue-green algae even though they're not at all related to plants, were very important organisms in the ancient world. In a real sense, we may owe our very existence to the evolutionary development of these lowly mats of slime billions of years ago.

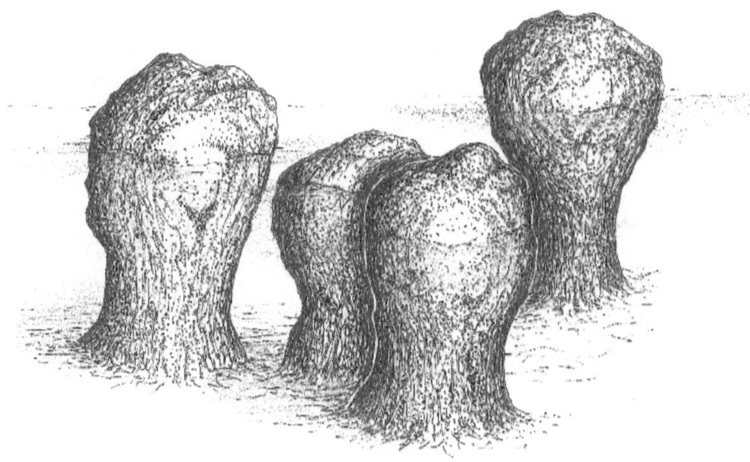

Stromatolites on a shallow seafloor

The deceptively primitive cyanobacteria developed an extraordinary skill early in their history. They learned how to photosynthesize – to take sunlight and convert it into chemical energy. Today, photosynthesis is commonplace, a skill shared by every green plant growing on Earth, but without it we would not be able to breathe. Photosynthesis utilizes carbon dioxide gas (CO_2), water (H_2O), and the energy from sunlight to create simple carbohydrates like the sugar glucose ($C_6H_{12}O_6$) with a very important waste product – oxygen gas (O_2). If you'd prefer a balanced chemical equation:

$$6\ CO_2 + 6\ H_2O \rightarrow C_6H_{12}O_6 + 6\ O_2$$

Water, of course, was easy for cyanobacteria to come by since they lived on the seafloor of shallow seas (shallow because sunlight is required for photosynthesis to work). What about the carbon dioxide gas? As we've already discussed carbon dioxide was much more abundant in the early Earth's atmosphere (and hence dissolved in the early Earth's oceans) than it is today.

The glucose that results from photosynthesis is used for energy in both plant and animal cells – we all have it circulating in our bloodstreams. While animals obtain glucose from the foods they eat, plants create it through the process of photosynthesis (which is why plants are called primary producers by biologists). The oxygen gas produced by photosynthesis is a waste product and is released into the environment by the organism just as we expel carbon dioxide, a waste product of respiration, when we exhale. Imagine a seafloor covered with mounds of green slime with little bubbles of oxygen gas rising off the cyanobacterial mats. This process didn't occur for a few years, a few millennia, or even a few million years. This process has been chugging along for well over three billion years now! This is one of the main reasons why our atmosphere today is 21% oxygen and only 0.04% carbon dioxide (78% of our atmosphere is nitrogen which has built up from volcanic outgassing over geologic time since it's relatively unreactive in our atmosphere). We are, of course, working hard to increase that percentage of carbon dioxide as we release tens of billions of metric tons of this greenhouse gas each year by the burning of fossil fuels – essentially reversing the photosynthetic process. Climate change skeptics are simply incorrect when they argue that humans can't affect the atmosphere of the Earth. The planet doesn't just affect life, life also affects the planet. The actions of organisms, the humble cyanobacteria as well as arrogant humans, can dramatically and permanently change the Earth's environment.

The stromatolite mounds built by cyanobacteria were important in the ancient world but are relegated to mere curiosities today. Stromatolites dominated the shallow seas for billions of years before the evolutionary development of a fierce predator – the voracious marine snail. Stromatolites were helpless against the ravaging attacks of these abundant mollusks and, after the evolutionary appearance of snails and other marine invertebrates, stromatolites became rare in the fossil record. They only exist today in harsh environments where they can eke out a living away from the cutthroat competition found among the modern marine invertebrates. One famous location goes by the intimidating name of Shark Bay in Western Australia, a hypersaline lagoon rife with poisonous sea snakes where the stromatolites exist as true living fossils of an earlier age.

While we've already discussed how the earliest minerals preserved in ancient sedimentary rocks are never oxidized due to the lack of free oxygen in the atmosphere, there is another type of rock which records gradually increasing oxygen levels on the ancient Earth. These rocks, called banded iron formations or BIFs, have been intensively studied due to their economic importance – they're one of the major sources of iron ore around the Earth. Banded irons generally formed between about 3.0 and 1.8 billion years ago and represent a time of low (but rising) oxygen levels on the Earth. They are unable to form if there is no free oxygen available, but are also unable to form under the present-day atmospheric conditions. While the exact mechanism by which BIFs formed is still subject to some debate, the key to their formation is that iron is only soluble in an acidic ocean. As discussed earlier, the oceans were originally acidic due to the high CO_2 concentrations in the Earth's earliest atmosphere. The iron in the BIFs is in the form of hematite, or iron oxide (Fe_2O_3), so there was free oxygen available; it just wasn't plentiful enough to alter the ocean's chemistry to the alkaline state that we see today.

Geologists refer to the time a little over 2 billion years ago as the "oxygen revolution." This is the time when oxygen levels built up high enough in the Earth's atmosphere such that life on Earth was forced to make adaptations to the changing environment. This is a common theme in the evolution of life – evolutionary changes often follow climatic change due to the Darwinian stresses of natural selection. It's hard to underestimate the importance of this event. Some geologists refer to it as the "oxygen crisis" or the "oxygen catastrophe" to emphasize the fact that this was a traumatic event for previously anaerobic life on Earth.

The biological consequence of this oxygen crisis was the first appearance of eukaryotes – organisms with a nucleus and organelles within their cells. The earliest eukaryotes were single-celled, just like the prokaryotes, only larger and more complex. The importance of eukaryotes is underscored by the fact that all multicellular organisms, including humans, belong to this large domain of life. Eukaryotes developed more complex metabolic processes to exploit the free oxygen that was now available in the Earth's environment. In animal cells, for example, there are organelles called mitochondria which utilize oxygen to create energy for the cell. Since mitochondria have their own genetic material, distinct from that in the nucleus of the cell in which they reside, biologists believe they arose through a process called endosymbiosis. Mitochondria were once free-living prokaryotes that "moved into" eukaryotic cells – a symbiotic arrangement that benefitted both organisms.

Eukaryotes are significant in another important way as well. Eukaryotes developed a new and novel method of reproduction that eventually resulted in an increase in the diversity of life on Earth. Prokaryotes, like bacteria, reproduce by binary fission. The genetic material in the cell makes a copy of itself, the original and the copy move to opposite ends of the cell, and the cell pinches and divides in two. Each resultant bacterium is genetically identical to the other – they are clones. In such a process, variability can only really occur when there's a change in the DNA through a mutation or copying error. Eukaryotes, on the other hand, developed sexual reproduction. In sexual reproduction, two organisms each contribute half of their genetic material to their offspring. The offspring are now genetically different from either of the two parents. This process introduces a lot more variability into a population of organisms. Why is this a good thing? Variability benefits a species because if conditions change, as they always do on planet Earth, some individuals may be better adapted than others to survive. English biologist Herbert Spencer referred to this as "survival of the fittest" in 1864 after reading Charles Darwin's *On the Origin of Species*. So, while sexual reproduction is a more complex process, it allows for natural selection to act upon individual genetic traits rather than the entire genetic makeup of an organism and eventually brought forth the amazing diversity of living things we see today.

We are now in the Proterozoic Eon, about 1.5 billion years ago. The area we know as New York State did not yet exist but the Earth had been evolving for the past 3 billion years. Two thirds of Earth history has passed us by! Oxygen levels are increasing in the atmosphere and the ocean's chemistry is gradually becoming more like that of the modern-day world. Life exists in the oceans – free swimming bacteria, stromatolite mounds on the seafloor, and the earliest eukaryotic cells – but the continents are still barren wastelands of rock. The stage is now set for the grand entrance of the rocks of the Hudson Valley region.

2
Rodinia

Mountains are Earth's undecaying monuments.

Nathaniel Hawthorne, *Sketches from Memory*

Despite the conceit of historians, the history of the Hudson Valley did not begin with Henry Hudson sailing his three-masted ship, the *Halve Maen* (Half Moon), up "the River of Mountains" in the autumn of 1609. Nor did it begin with the arrival of the woodland Paleo-Indians a few thousand years earlier. As a matter of fact, if we could compress the entire history of the Earth down to a 24-hour span of time, all 6,000 or so years of recorded human history only fits into the last tenth of a second of that day. The true history of the Hudson Valley began some 1.5 billion years earlier when proto-North America, known to geologists as Laurentia, was deep in the Southern Hemisphere.

While it sounds odd to talk about North America being in the Southern Hemisphere, modern geology has shown that the rigid outer layer of the Earth, called the lithosphere, is split into a dozen or more tectonic plates. These plates are constantly moving and jostling against one another, driven by convecting heat deep within the Earth, at the yawn-inspiring rate of a few centimeters per year. This process is referred to as plate tectonics (*tekton* is the Greek word for "builder") and is the mechanism by which continents move around the surface of the Earth over the long spans of geologic time. As we'll soon see, it's also how all of the Earth's mountain belts were created.

Like an errant child wandering away from their parent, North America is currently moving away from Europe at a rate of about 2.5 centimeters per year, ever widening the Atlantic Ocean between us. Assuming this rate has been constant (not always a valid assumption), and knowing that Europe and North America are presently about 5,000 km apart, we can use simple math (time = distance / rate) to determine that the Atlantic Ocean didn't even exist some 200,000,000 years ago (which is exactly what geologists claim). While 200 million years sounds like a long time to most people, it's not to a geologist. Geologists are used to thinking in terms of "deep time," and this inconceivably vast span is only 4% of the Earth's 4.5 billion-year history. Even a snail, if given enough time, can completely circumnavigate the globe.

There are two complementary themes which will constantly recur within this book – plate tectonics and geologic time. The Earth's internal heat engine has been driving tectonic plates for most of Earth history with dramatic consequences played out over incomprehensibly long spans of time. Plate tectonic collisions build mountains; and Himalayan-scale mountains have formed and then completely eroded away here in the Northeast. Not once, but several times. Geologists call these mountain-building events orogenies (from *oros*, the Greek word for mountains) and the first such event to affect our area is known as the Grenville Orogeny (named after a village in Québec where rocks affected by this event were once studied).

Plate tectonics is a relatively recent concept in geology. While speculation about continents drifting across oceans has a long history, the German scientist Alfred Wegener is generally credited with first proposing scientific evidence for this hypothesis in a book called *The Origin of Continents and Oceans* published in 1915. Wegener's revolutionary idea was ahead of its time and generally rejected by geologists of the day due to the lack of a plausible mechanism to explain how masses of continental crust are able to plow through the basaltic oceanic crust. This inability to come up with a viable mechanism is not surprising given how little science knew about the nature of the deep ocean seafloor and the interior of the Earth in the early 1900s.

Throughout the 20th century, new technologies developed which allowed us to sample and map the seafloor, measure changes in the ancient magnetic field of the Earth (which are preserved in rocks), and to image the interior of the Earth from the passage of seismic earthquake waves. This data gradually led to the reformulation of Wegener's original idea of continental drift into the modern idea of plate tectonics, primarily in the 1960s and 70s.

As mentioned, plate tectonics is the concept that the rigid outer layer of the Earth is split into a dozen or more tectonic plates. This outer layer is called the lithosphere (*lithos* is the Greek word for "stone") and is composed of both oceanic and continental crust as well as the uppermost layer of the mantle – all of which behave as a rigid solid. These lithospheric plates are gliding over a softer layer of the mantle called the asthenosphere (*asthenos* is also derived from Greek and means "without strength").

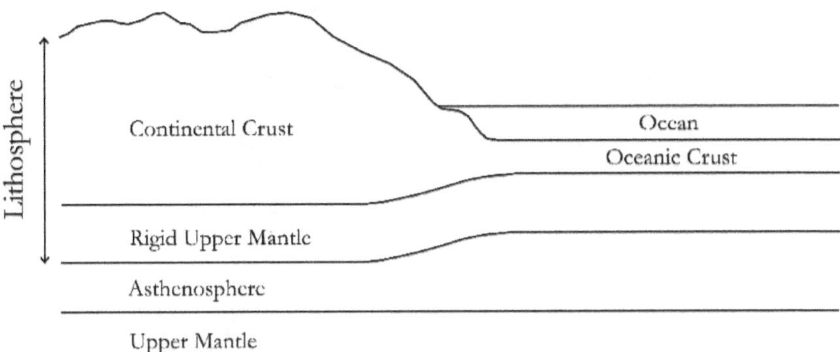

Crust and mantle cross section

In other words, it's not just continents that move, plates move and plates are composed of both continental and oceanic crust as well as the uppermost part of the mantle.

The driving force behind these plate movements is the convective circulation of material within the deeper mantle (below the lithosphere). Hot mantle material rises since it's less dense while cooler mantle material sinks due to it being more dense. This up and down movement within the mantle forms convection cells driven by heat energy from the interior of the Earth. If you're old enough to remember, think of a lava lamp. For those who would prefer a less esoteric example, a pot of bubbling oatmeal on a hot stove illustrates the same concept. One main difference, however, is that the mantle is actually solid due to the high pressures within the Earth's interior. Yes, solids can flow as well as liquids – it simply takes longer amounts of time (one thing we have plenty of in geology).

While controversial when the theory of plate tectonics was first being developed, plate movements today can be directly measured with radio signals from global positioning system (GPS) satellites orbiting the Earth and the theory is now ensconced as a central unifying concept in modern geology. While geologists still argue about the exact mechanisms driving plate movements and the exact configuration of plates in the geologic

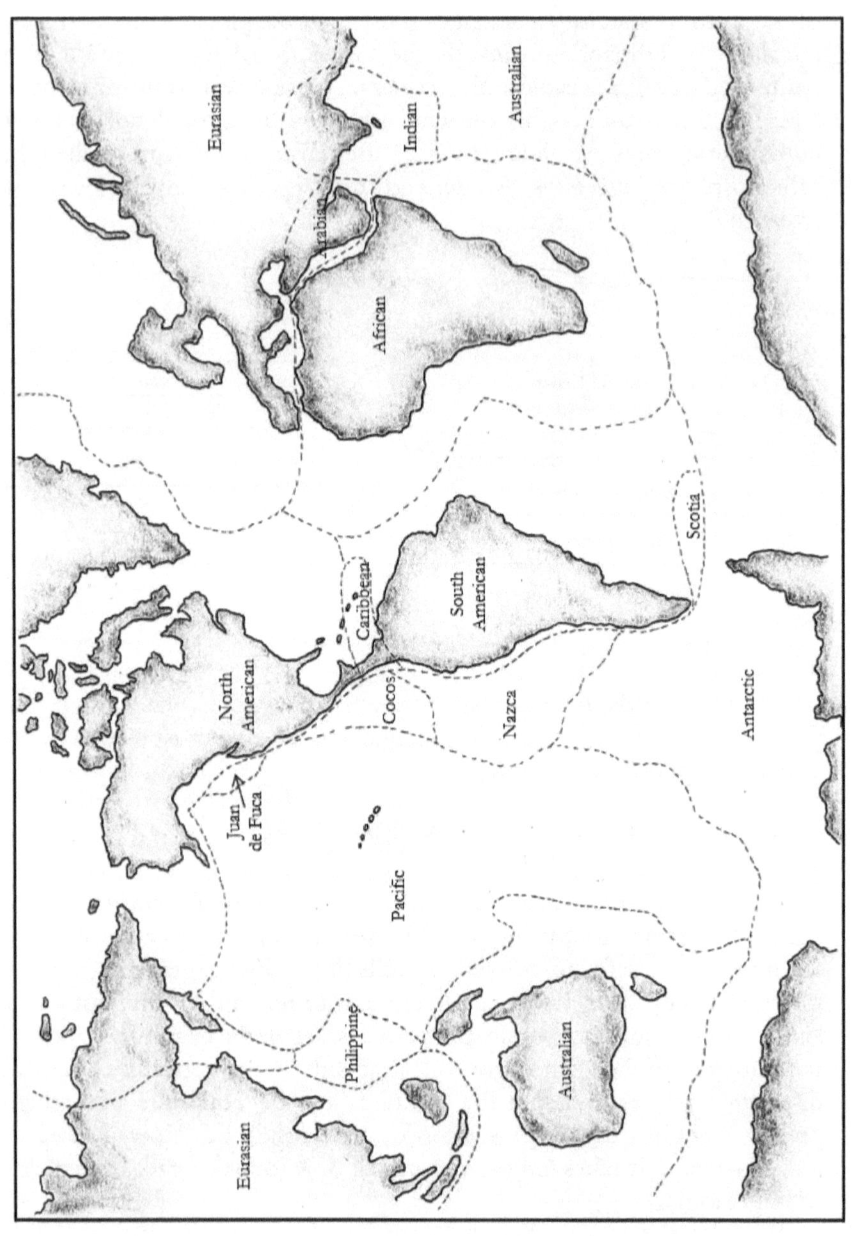

The Earth's tectonic plates

past, the fact remains that tectonic plates have moved over time and are still moving today. As we'll see later, without this ceaseless movement the Earth would likely be a lifeless, dead planet.

Back to the area that is poised to become New York State. The time is now 1.5 billion years ago – a time geologists call the Proterozoic Eon. Proterozoic literally means "former life" referring to the mostly microscopic life which existed back then. If we could travel back to this Eon, we'd see an Earth that looked very different from the one we're used to today. We would see a number of continents arranged in a global ocean but none would be familiar in size, shape, or position. The continents would be rocky and barren, devoid of plant and animal life, with only the noise of the wind and surf breaking the silence. The oceans would have life, but most of it would be microscopic and invisible to the human eye. The only readily apparent life would be small mounds in the shallow sunlit water covered with a greenish slime – the stromatolites discussed earlier that were tirelessly pumping increasing amounts of oxygen into the Earth's atmosphere.

The stage is now set for the formation of the ancient crust which underlies the Hudson Valley (and much of the eastern seaboard). Around 1.5 billion years ago, the area that's now eastern New York was the sediment-covered floor of an ancient sea just off the southern coast of Laurentia or Proto-North America. What we know today as the continent of North America was much smaller then and oriented differently than it is today. Laurentia was rotated 90° from present-day North America and straddled the equator. In other words, the equator ran through what is now central Canada, the Midwestern United States, and Mexico.

In the oceans surrounding Laurentia were other barren blocks of continental crust. The continental block closest to the southern coast of Laurentia (which is now the east coast of North America – confused yet?) was thought to be Amazonia. Today, Amazonia is one of several pieces of crust which comprise the modern-day continent of South America but during the Proterozoic Eon it was drifting through the southern seas by itself. If there had been anyone around at the time to observe the landscape, they might have noticed that Amazonia was a few centimeters closer to Laurentia each year.

How do blocks of continental crust approach each other? Remember that plate tectonic theory tells us that continents don't move by plowing through stationary oceanic crust. Both continental and oceanic crust move together as a tectonic plate. If continents are approaching each other, something has to happen to the intervening oceanic crust. That something is a process called subduction.

As tectonic plates move on the surface of the Earth, there are three basic types of interactions which can occur at plate boundaries. Plates can pull away from each other (a divergent margin), plates can move toward each other (a convergent margin), or plates can move sideways past each other (a transform margin). In reality, transform margins can exist simultaneously with convergent or divergent boundaries (imagine pushing on something at an angle) but we'll ignore that complication for now. These plate boundaries are where all the action is, geologically-speaking, since pushing or pulling on a rigid tectonic plate forces it to do interesting things.

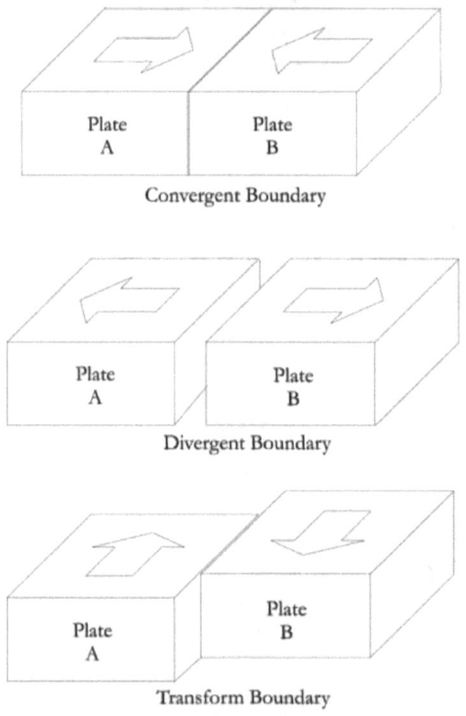

The three major types of plate boundaries

Laurentia and Amazonia were approaching each other because mantle circulation beneath their two plates was inducing a compressive stress – in other words, they were being squeezed together. To accommodate this compressive stress, something had to give and what gave was the thinner oceanic crust of the seafloor. The seafloor fractured and started to slide beneath the thicker continental crust of Laurentia. This process is called subduction and leads to some dramatic consequences. As the oceanic crust subducts, it forms a deep trench on the seafloor. As the tectonic plate descends into the mantle, it doesn't slide down smoothly. It slides in what's called a "stick-slip" motion where the plate gets stuck, stresses build up, and then portions of the plate suddenly slip. Each slip generates an earthquake and some of the largest earthquakes on Earth are generated along slabs of subducting oceanic crust. The pressure of plate convergence also results in folding, faulting, and uplift of the rocks comprising the continental crust next to the subduction zone. This deformation thickens the crust forming a mountain belt along the coastline of the continent.

As the plate descends deeper into the mantle, it initiates melting – typically below 100 kilometers of depth (that's a little over 60 miles down). This melting generates molten rock, or magma, and this magma, being less dense than the surrounding mantle material, will gradually work its way up toward the surface. Upon reaching the surface, magma erupts to form volcanoes. These volcanoes erupt through the folded and faulted mountain belt forming what geologists term a continental volcanic arc – a chain of volcanic mountains on a continent next to an offshore subduction zone.

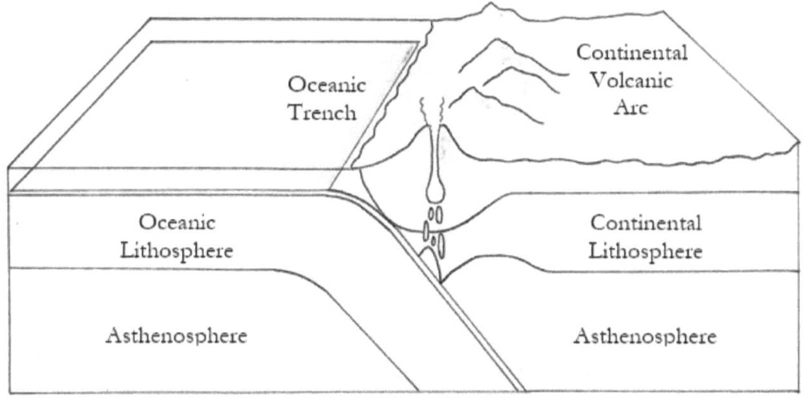

A continental volcanic arc next to a subducting oceanic plate

In our modern world, we see a good example of a continental volcanic arc in the Andes Mountains running along the western coast of South America. Just offshore is the Peru-Chile Trench where a small plate of oceanic crust, called the Nazca Plate, is subducting beneath South America. The Andes are highly folded mountains with numerous active volcanoes all along its spine. Earthquakes are also a frequent occurrence here; one of the largest earthquakes ever recorded by seismographs was a magnitude 9.5 just off the Chilean coast in May of 1960.

So, Amazonia approached Laurentia due to the consumption of the intervening oceanic crust down into a subduction zone. Earthquakes shook the landscape and a continental volcanic arc formed on Laurentia starting around 1.3 billion years ago. These large volcanoes periodically erupted and large amounts of volcanic ash settled into the surrounding seas. These eruptions were catastrophic events for the seafloor stromatolites (it's hard to photosynthesize when you're buried under sediment) and the ash was eventually incorporated into the rocks which later formed from these seafloor sediments. The problem with this scenario is that it simply can't continue forever (over 100 million years, yes, but that's not a huge amount of time in geology). The consumption of the seafloor meant that Amazonia grew ever closer to Laurentia and they eventually collided around 1.1 billion years ago.

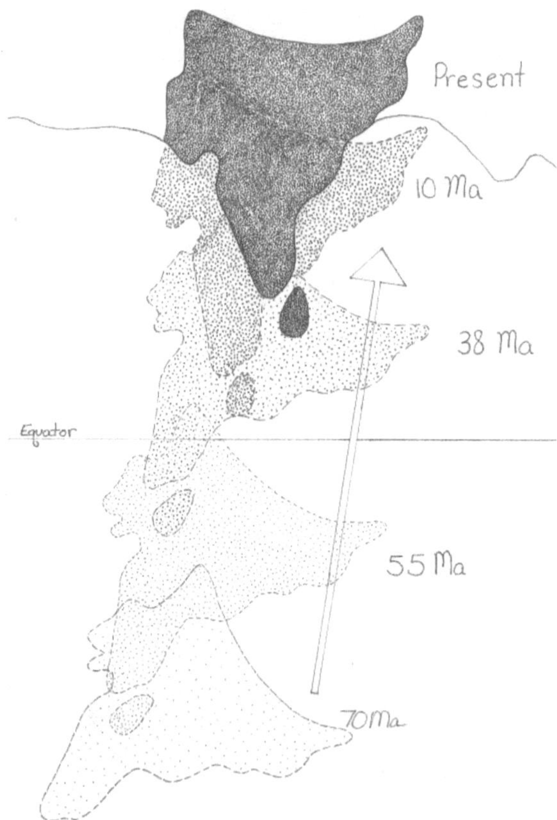

India approaching Asia from 70 million years ago until present

This 1.1 billion year old collision of Amazonia with Laurentia is analogous to a far more recent event. Approximately 80 million years ago, the subcontinent of India was in the southern Indian Ocean near the present-day African island of Madagascar. Over the next few tens of millions of years, India drifted northward toward the southern coast of Asia. As India moved northward, there was a small ocean between it and Asia. Geologists refer to this as the Tethys Sea. Limestone formed on the basaltic seafloor crust of the Tethys due to the large number of hard-shelled marine invertebrates living on the subtropical seafloor. Limestone formation is interesting in its own right and we'll be discussing it in more detail in the next chapter. Close to Asia was a deep seafloor trench where the Tethyan seafloor was being subducted beneath Asia. The part of Asia that is now Tibet was home to a continental volcanic arc due to rising magma generated by the melting seafloor plate as it subducted.

Starting about 50 million years ago, India collided. Subduction shut down and, as India began to push into Asia, the modern Himalayan Mountains were thrust up by deformation of the rock caught between the colliding plates. The Himalaya are still rising as India continues to shove into the underbelly of Asia. That's why there's marine limestone on the summit of Mount Everest – remember McPhee's earlier quotation in the Prelude? The marine limestone is part of the Tethys Sea which was caught up in the collision and raised to the roof of the world. Geologists can tell that this marine limestone was not originally deposited *in situ* in the high Himalaya (during a mythical global flood, for example), because the limestone is heavily folded and faulted and surrounded with other classic signs of mountain building such as remnants of volcanic arcs, igneous intrusions, and highly-metamorphosed rocks.

Continental collision is a slow-motion train wreck. Continental crust is too thick to subduct so two huge masses of granitic crust will simply grind into each other while crumpling and fracturing rock around the collision zone. Because the rock is shortening horizontally, it accommodates by thickening vertically resulting in a high mountainous area of heavily deformed rocks. Deep in the core of the mountain belt, high heat and pressure physically and chemically alter the rock in a process known as metamorphism. Deeper still, the heat is high enough to melt rock resulting in masses of magma which slowly rise through the crust. Most never make it to the surface and cool to form large intrusive bodies of granite deep underground. Since tectonic plates only move at a speed of a few centimeters per year, this orogenic (mountain building) process typically occurs over tens of millions of years. This Himalayan Orogeny, which is occurring as we speak, is essentially unnoticed by the people living in the area. All they observe on the scale of a human lifespan is an occasional large earthquake or landslide as the mountains attempt to accommodate the relentless stresses of plate convergence.

Around the same time that Amazonia was colliding with Laurentia, other pieces of continental crust were colliding elsewhere around the world. The period of time just over a billion years ago was special in Earth history as various blocks of continental crust collided, an event called the Grenville Orogeny in our part of the world, and accreted to form a massive supercontinent known as Rodinia (from the Russian word for "Motherland"). A reconstruction of Rodinia shows that the Laurentia-Amazonia boundary roughly corresponds to the present-day east coast of North America and that the block of crust that collided with what's now the western part of North America was thought to be Antarctica. The geography of the Proterozoic Earth was far different than that of our modern-day world.

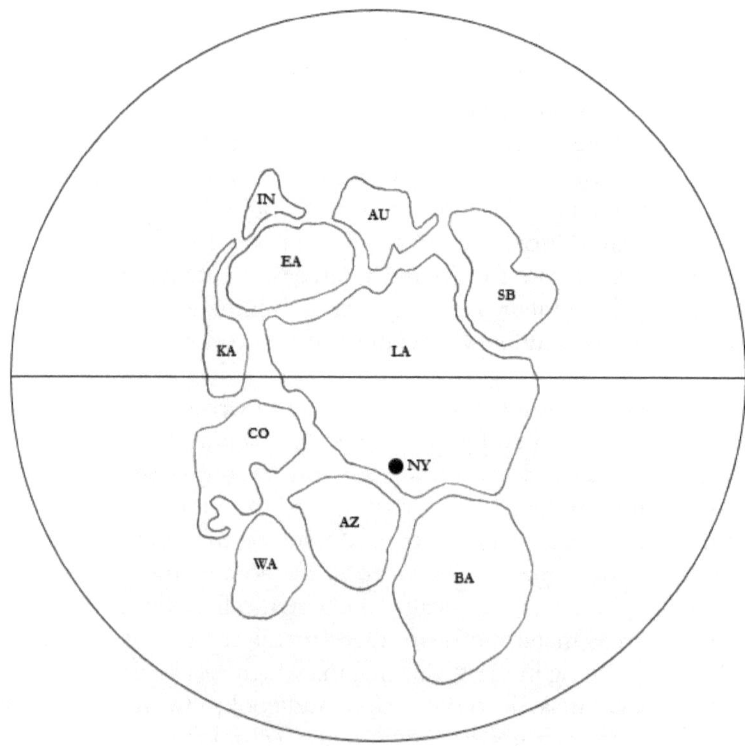

One possible simplified reconstruction of the Late Proterozoic Eon supercontinent of Rodinia. Location of New York shown in Laurentia (LA) or Proto-North America. Other continental blocks include Australia (AU), Amazonia (AZ), Baltica (BA), the Congo (CO), East Antarctica (EA), India (IN) the Kalahari (KA), Siberia (SB), and West Africa (WA).

The massive mountains that formed along the collision zones between these blocks of continental crust were Himalayan in scale and a portion of these Grenville Mountains between Laurentia and Amazonia ran right through what is now New York State. Grenville rocks in New York State date from approximately 1.3 to 1.0 billion years ago and record at least two different mountain building phases occurring over a span of some 300 million years. While we will refer to all of these events collectively as the Grenville Orogeny, keep in mind that geologists often talk amongst themselves about two distinct phases of this orogeny termed the Elzevirian (1.3-1.2 billion years ago) and the Ottawan (1.1-1.0 billion years ago). The Elzevirian appears to represent the formation of the subduction-related continental volcanic arc while the Ottawan represents the main phase of continent-continent collision between Laurentia and Amazonia.

Rocks that formed during the Grenville Orogeny can be traced today from the Canadian Maritimes down the eastern coast of the United States and then westward into Texas. Prior to the Grenville Orogeny, these areas of continental crust did not exist.

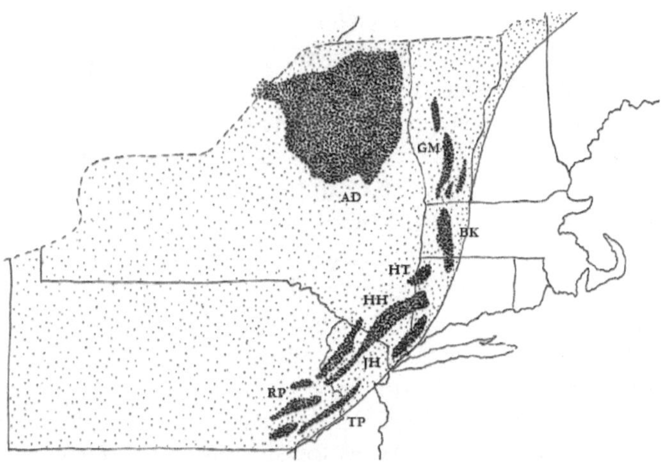

Light stipples indicates the buried extent of Grenville-aged rocks in the northeastern U.S. and dark stipples indicate surface outcrops. Symbols denote the Adirondacks (AD), Berkshires (BK), Green Mountains (GM), Hudson Highlands (HH), Housatonics (HT), Jersey Highlands (JH), Reading Prong (RP), and Trenton Prong (TP).

Since Grenville-age rocks form the continental crust of New York State, the New York State region did not exist prior to 1.3 billion years ago. This is the beginning of Hudson Valley geology. In most areas of

New York, the Grenville rocks form what are called the basement rocks since they are covered by several kilometers of younger sedimentary rocks. They're only exposed at the surface in two regions, the Adirondacks and the Hudson Highlands. Basement rocks are exposed in the Adirondacks and Hudson Highlands because those are areas where they were later uplifted and these ancient rocks were able to punch through the overlying cover strata of younger sedimentary rocks (events we'll discuss later).

Harriman State Park, in the heart of the Hudson Highlands, is an excellent area to study Grenville Province igneous and metamorphic rocks. I realize that the Hudson River begins in the Adirondacks, another large region of Grenville-age rocks, but much of what we'll learn about the more accessible Hudson Highlands will apply to the Adirondacks as well. The Highlands begin at the Berkshires along the New York/Connecticut border and run southwest across the Hudson into the New Jersey Highlands and then into an area of Pennsylvania called the Reading Prong.

The first clue as to the nature of the Highlands is obvious after even a casual glance at a map of the Hudson River. North of Storm King Mountain, near Newburgh, the Hudson is relatively wide. As you move south through the Highlands, however, the Hudson River becomes very narrow and deep. Moving out of the Hudson Highlands into Haverstraw Bay and then the Tappan Zee, the Hudson once again becomes a wider, shallower river. This difference in the width of the river is due to the nature of the rocks in the Highlands – they are hard and resistant igneous and metamorphic rocks. North and south of the Highlands one finds relatively soft and easily eroded sedimentary rocks such as shale. Observing the landscape for clues to the underlying geology is a skill developed through practice and falls into the realm of geology termed geomorphology (literally "Earth form").

The geology of the Hudson Highlands is very complex in detail and still an area of active research by geologists. Basically, the region consists of high-grade metamorphic rocks intruded by masses of igneous rock. Metamorphic rocks form when preexisting rocks are altered by heat and pressure. High-grade metamorphic rocks were altered by near-melting-point temperatures around 600-800° C (over 1,100° F) and pressures of 400-700 megapascals (up to 100,000 pounds per square inch). These extreme temperature and pressure conditions correspond to depths of about 14-25 kilometers (over 9 miles below the surface). Igneous rocks form when high enough temperatures are reached to produce magma; partial melting generally begins for most rocks at temperatures above 800° C (1,472° F) with different minerals melting at different temperatures. When magma cools and crystallizes, an igneous rock results. Igneous intrusions form because magma is mobile, less dense than the surrounding solid rock, and can therefore travel upward through fractures in the crust to solidify within bodies of preexisting rock (in this case, within the metamorphic rocks of the Highlands). Where would rocks experience such high temperatures and pressures? In the case of the Hudson Highlands, these hellish conditions were reached kilometers below the surface in the core of an ancient mountain belt – the Himalayan-sized Grenville Mountains. This is an important concept in geology; rocks generally form deep within the crust of the Earth, not at the surface. Many of the rocks we walk on today were once deeply buried and are only exposed at the surface after long periods of erosion of the overlying material.

Enough talking, as I would tell my geology students, let's go look at rocks. Throughout this book I will be describing places in the Hudson Valley where you can easily see examples of the geology we'll be discussing. I encourage you to visit these locales and examine the rocks in person. Only then can you gain a real appreciation of how geologists decipher the long history of the Earth. Besides which, most of these places are beautiful, interesting in their own right, and well worth a visit.

One place to easily observe Grenville-age rocks and view the Hudson flowing through the Highlands is from the top of Bear Mountain in Harriman State Park. There are several places to observe the Hudson River from Bear Mountain but the view from many of the road pull-offs is actually better than the panorama from the observation tower at the top of the mountain. The narrowing of the Hudson River as it flows through the Highlands can clearly be seen. Distinctive nearby peaks are Anthony's Nose on the opposite side of the Bear Mountain Bridge across the Hudson and the prominent ridge of Dunderberg Mountain to the south.

The rock exposed on the top of Bear Mountain is called the Storm King Granitic Gneiss by geologists. As you could guess, this rock also crops out on its namesake, the top of Storm King Mountain to the north. Hike around a bit and examine this rock – it has some interesting features. This rock began its life as granite, an intrusive igneous rock, and was later metamorphosed into gneiss, a high-grade metamorphic rock. This is not uncommon in mountain belts; melting forms igneous rocks which are then metamorphosed by the high temperatures and pressures from continued mountain building. Gneiss is easily recognizable since it has a metamorphic foliation – a pervasive layering within the rock defined by layers of lighter- and darker-colored minerals. The lighter-colored layers are formed from quartz and feldspar (quartzofeldspathic) minerals and the darker-colored layers are formed from minerals rich in iron and magnesium (ferromagnesians). This so-called gneissic layering forms due to the intense pressure which occurs during mountain building.

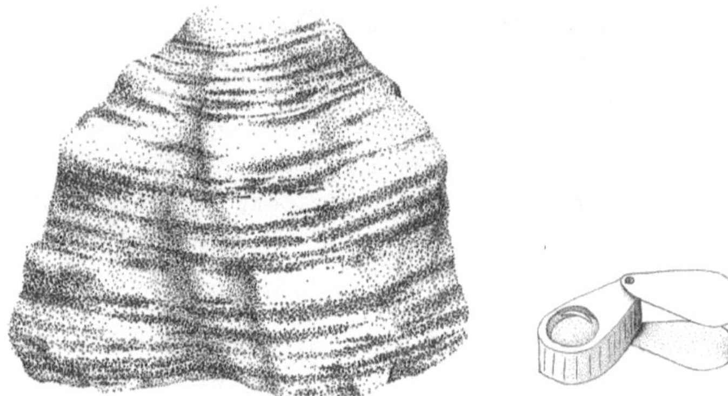

Gneissic layering and a geologist's hand lens

From the degree of metamorphism and the radioactive dating of elements within the rock's minerals, geologists can determine that this granitic gneiss formed as deeply as 40 kilometers (25 miles) below the surface just over a billion years ago. Remember that the reason we can now walk on rocks that were once buried kilometers below the surface of the Earth is because those many kilometers of overlying rock have since eroded away into sediments. Walk around a bit on the exposed rocks and look carefully at the little hollows and depressions in the granitic gneiss. Notice the sediments? Look more carefully. What minerals do you think compose these sediments? If you examine the sediment with some magnification (geologists routinely carry around a small, high-quality 10x magnification hand lens), you'll notice that the sediment grains are primarily glassy grains of quartz (with a few grains of associated darker minerals). This isn't at all surprising to a geologist. Quartz sediments are ubiquitous in our world – much of what we know as sand is, in fact, quartz eroded from ancient rocks (other minerals often weather into finer clays which are simply washed away).

Analysis of the bulk composition of the Earth's crust shows that the two most abundant elements are oxygen (O) and silicon (Si) at 46.6% and 27.7% by weight. That's because most of the Earth's crust is composed of what geologists call silicate minerals (they're also known as the rock-forming minerals). Silicates are minerals with both silicon and oxygen in their formula along with other common elements such as iron, magnesium, calcium, potassium, and sodium. The simplest silicate mineral is quartz with the formula SiO_2 – hence its abundance in nature. Quartz is also extremely resistant to both physical and chemical weathering. It's harder than steel and mostly impervious to acids. These properties ensure that when quartz-rich rocks like granitic gneiss weather, quartz sediment grains are often the only thing left behind. While we see this on a small scale in the present-day Hudson Highlands (small grains of quartz sand from the eroding gneiss), we will see it on a much larger scale later in our story.

There are many other igneous and metamorphic rocks exposed in the Hudson Highlands besides the Storm King Granitic Gneiss. Driving down from Bear Mountain and exploring the surrounding area in Harriman State Park will reveal an assortment of rock types representing different parts of the Grenville mountain building event. In some areas, outcrops of unmetamorphosed granite can be seen. This granite is very similar in appearance to the granitic gneiss but it's not foliated since it did not experience significant amounts of metamorphism. This means that the mineral grains within the granite are not segregated into light and dark layers (as they are in a gneiss) but are instead randomly oriented throughout the rock. These masses of granites were intruded as magma during mountain building and generally date from the period between 1.1-1.0 billion years ago.

Outcrops of distinctive rusty-red gneiss can also be seen throughout Harriman State Park and the surrounding area. This is the same basic type of rock as the Storm King Granitic Gneiss, in that they're both high-grade metamorphic rocks, but they had very different origins. The rusty-red gneisses seen throughout the area are not metamorphosed granites. They began their lives as sedimentary rocks. Detailed analysis of the minerals and geochemistry of these gneisses reveals that they originated as volcanic sediments on a seafloor. These sediments were apparently derived from two source regions; some were from erosion of the continental volcanic arc on Laurentia while others were derived from erosion of continental rocks exposed on the microcontinent of Amazonia. These sediments then washed down into the shrinking sea between these two small continents. This was, of course, prior to their collision so layers of volcanic ash were also deposited by periodic eruptions of the nearby volcanoes. These rocks contain a lot of sulfur which typically indicates deposition in an oxygen poor environment – in other words, a deep marine basin near the subduction zone with poor water circulation and oxygen-poor bottom conditions (a euxinic basin). These sediments became metamorphosed during the collision which formed the Grenville Mountains.

While many other specific igneous and metamorphic rocks occur in the Highlands, they're really all variations on a theme. Sediments, some volcanically derived, accumulated on the seafloor between Laurentia and Amazonia and were subsequently caught up in the continental collision and metamorphosed. High heat within the core of the mountain belt formed magmas which ascended and cooled into igneous intrusions. Some of these igneous intrusions were later metamorphosed during continued mountain building. The bottom line is that by a billion years ago, Amazonia had collided with Laurentia to help form the supercontinent of Rodinia. In what is now New York State, the massive Grenville Mountains were formed from a belt of igneous and metamorphic rocks. No sooner had the mountains formed, however, that their inevitable destruction ensued.

In the modern Himalaya, compressional mountain building from the northward movement of India into Asia is associated with lateral movement of the crust along large east-west trending faults in Tibet. In other words, as India pushes into Asia, blocks of crust are fracturing and squeezing out sideways (a fracture with movement on either side is termed a fault). This is sometimes called escape tectonics and occurs during large-scale continental collisions. These escape tectonic features have also been identified in the Hudson Highlands. Since the rocks we see today in the Highlands were once much more deeply buried, however, these structures are not faults (brittle fractures in the rock along which movement occurred) but rather shear zones.

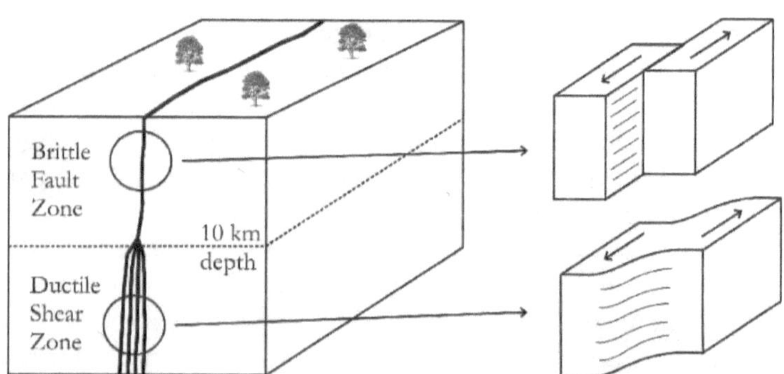

Zones of brittle faulting and ductile shear zone formation in the crust

To understand shear zones, we need to understand how rocks behave at depth. At the surface, when rocks are subjected to high amounts of stress, they will simply fracture and break. This is called brittle deformation. As we move down into the crust, however, temperatures and pressures increase such that rock begins to behave differently. When we reach depths of around 10 km (6 miles) or so, rock will start to flow when it's subjected to stress. This is called ductile deformation and it's how folds form in rocks. Many of the rocks in the Hudson Highlands clearly show folding. This is another indicator, along with the high degree of metamorphism, of their original depth of burial.

During the latest phase of Grenville Orogeny mountain building, large faults developed in response to escape tectonics. As these large faults extended downward into the crust, they turned into shear zones – narrow zones of ductile deformation. A shear zone is like a fault in that it allows two blocks of rock to slide past each other. Instead of sliding along a fracture between the blocks, however, a narrow zone of rock accommodates movement by flowing (while still in a solid state). If you would like to model this process, get two small pieces of rock and squeeze some modeling clay between them. You can now slide one rock past the other and they will move on the thin deforming layer of soft clay. Shear zones are recognizable today in a number of areas of the Hudson Highlands and studies have shown that there may have been hundreds to thousands of kilometers of offset along these large, regional-scale structures.

One very significant event associated with movement along these shear zones was the regional emplacement of iron ore minerals. It's possible to assign a radiometric age to movement on these shear zones and they appear to have formed primarily between 900 million and 1 billion (1,000 million) years ago. This is after the main phase of mountain building as things were starting to cool down. This cooling allowed some of the shear zones to move across the ductile/brittle transition and become open fractures – in other words, as the rock cooled, rocks became too hard to flow and instead fractured.

Water is ubiquitous in the Earth's crust. In areas of active mountain building, water will flow through the crust, heat up, and dissolve elements from the surrounding rocks. As this mineral-rich water flows through fractures and faults within the rocks, it will eventually cool and precipitate minerals within the fractures. Geologists refer to these mineral-filled fractures as veins. Most veins are composed of common minerals like quartz (SiO_2). In some mountain belts, such as the Rockies or Sierra Nevada Mountains out west, this process also led to the precipitation of gold and silver veins. The Hudson Highlands, however, was blessed primarily with veins of iron associated with the dilated shear zones.

The iron mineral formed in the Highlands was mostly magnetite, a form of iron oxide (Fe_3O_4). Prospecting for the narrow veins of magnetite is relatively easy; simply walk around with a compass and note where the needle is deflected by the iron ore. If you feel the need to confirm the identity of the heavy, shiny black mineral, check to see that it attracts a magnet (hence the name magnetite).

Magnetite iron ore from the Hudson Highlands was very important during the first hundred years of our country's history. Hiking in Harriman State Park today will lead you to a number of abandoned iron mines — some of which had extensive underground workings which are no longer accessible having naturally filled in with groundwater. Surrounding many of the mines are piles of black rock composed mostly of magnetite.

The discovery of iron led to the development of these mines as well as towns, roads, furnaces, and foundries — most of which today are mere ruins hidden within the forests of Harriman State Park and the surrounding area. Magnetite is relatively easy to smelt — simply heat up the mineral with some charcoal (a source of carbon) to drive off the oxygen in the form of carbon dioxide (CO_2) gas leaving behind the molten iron. In the late 1700s and into the 1800s, forests in the Highlands were clear-cut to provide charcoal for heating the iron ore (and farmland to support the communities which arose around the numerous mines and furnaces). The charcoal was loaded into furnaces with the iron ore and running water from streams powered the bellows which kept the furnaces burning. As molten iron flowed out of the bottom of the furnace, it was cast into sand molds forming ingots called "pigs."

The Hudson Highlands and its iron ore played a strategic role in the history of our country. During the Revolutionary War, the British sought to control the Hudson River in an effort to cut the colonies in half. Since the Hudson River narrowed as it flowed through the Highlands, this area became an important military chokepoint and led to the construction of a number of forts including Forts Clinton and Montgomery at the base of Bear Mountain and West Point a bit further upriver near Storm King Mountain. Hudson Highlands iron ore was used to forge the great chains that were strung across the Hudson at Fort Montgomery in 1777 and West Point in 1778. Remnants of this latter chain can still be seen at the United States Military Academy at West Point.

The Hudson Highlands iron industry remained strong throughout the 19th century until the discovery of massive amounts of high-quality iron ore in the Mesabi Range of northern Minnesota in the late 1800s. While there's still plenty of iron ore left within the Hudson Highlands, it's simply no longer economically, politically, or ecologically feasible to mine.

Back to the Proterozoic Eon circa 900 million years ago. The snow-capped Grenville Mountains would have been an awe-inspiring sight rising majestically from the plains of Rodinia. Hawthorne had it wrong in this chapter's epigraph, however, when he wrote that "Mountains are Earth's undecaying monuments." Mountains are awe-inspiring and appear immortal to those of us only allotted three score and ten, a few more if we're lucky, but nothing lasts forever and streams and rivers running off the Grenville Mountains would have gradually eroded them away. This is analogous to the present-day erosion of the Himalayan mountains as streams and rivers ceaselessly transport its sediments into the Ganges River and eventually down to the Bay of Bengal.

Let's suppose the Grenville Mountains eroded at a rate of one-tenth of a millimeter per year (0.1 mm/yr). This is not an unreasonable assumption since studies have shown that some areas of the modern high Himalaya have erosion rates as high as 10 mm/yr. If that rate is constant (which it never is in reality but it's a good starting assumption), how long would it take these Himalayan-scale mountains to completely erode away? The height of Mount Everest, the highest in the world, is approximately 8,850 meters (29,000 feet). Once again, we can use a simple equation (time = distance / rate) to determine that Everest-height mountains could completely erode away in only 88,500,000 years. Leave it to a geologist (or an economist) to use the terms "only" and "88 million" in the same sentence. While 88 million years seems like a long time to us, it's a geological drop in the bucket (the concept of "deep time" yet again).

The erosion of the Grenville Mountains most likely took longer than 88 million years as there are many complications to this simple picture. Rates of erosion in the Himalaya depend upon factors like rainfall amounts (areas subjected to India's monsoonal rains have higher rates of erosion), temperature variations (more physical weathering will occur in areas where the temperature fluctuates above and below the freezing point of water due to freeze-thaw cycles), and vegetation cover (which wasn't a factor in the ancient Grenville Mountains since land plants didn't yet exist). There's also evidence that the Grenville Orogeny was complex with several pulses of mountain building separated by periods of erosion. Despite our difficulty in working out an exact chronology of events a billion years ago, we do know that the Grenville Mountains were mostly gone by 600 million years ago (half a billion years after they formed). The area once crowned with high mountains was now a relatively flat, featureless plain.

3
Passive Margin

Nothing is world-wide, but everything is episodic. In other words, the history of any one part of the earth, like the life of a soldier, consists of long periods of boredom and short periods of terror.

Derek V. Ager, *The Nature of the Stratigraphical Record*

The only constant in the history of the Earth is change. Sometimes the change is relatively abrupt and catastrophic, but usually it's slow and gradual. This is the type of change we see only in hindsight, not while it's occurring. The Earth is changing as you read this. Mountains are growing and mountains are eroding. The rates at which this occurs, however, is so slow that it's essentially unnoticeable on the scale of a human lifespan.

Time is fundamental in geology. When looking at a rock, it's only natural to wonder about its age. Learning about its age also tells us something very important about the geological history of the area. Marine fossils in limestones in the Hudson Valley, for example, tell us that our area was once covered by a shallow, subtropical sea. The next logical question would be: "When?" Since we find marine limestones all over the world, even high in the Himalayan mountains, couldn't this provide support for a past global flood such as that recounted in the book of Genesis?

While limestones and other marine rocks around the world might have supported the idea of a global flood if they were all of the same age, it turns out that they're not even close. Even worse, layers of rocks deposited in a marine environment are often separated by other rock strata that were clearly deposited in a terrestrial environment. It did take a while for all this to be figured out, and not without a fight from supporters of the traditional Biblical worldview, but by the late 1700s, naturalists were starting to argue for an ancient Earth with a long history of rock formation.

The problem faced by these early geologists was that radiometric dating didn't yet exist. How could they tell the age of rocks? And even if radiometric dating did exist, one couldn't use it to date sedimentary rocks (still generally true today). The reason for this is that radioactive isotopes can move into and out of minerals within rocks while they're hot through a process called diffusion. When the temperature of a specific mineral within the rock cools below a certain temperature, called the closure temperature, the radioisotopes are then locked into the rock and the radioactive decay clock essentially begins for that rock. It makes sense that we can date igneous rocks, which cool from magma, and metamorphic rocks, which are heated to high temperatures and then cool again, but what about sedimentary rocks?

Imagine a sandstone – a sedimentary rock composed of quartz sand grains cemented together. Sandstones commonly contain small amounts of minerals other than quartz and many of these other minerals contain radioisotopes. Let's pull one of these minerals out of the sandstone and obtain a radiometric age. Is this the age of the sandstone? No, it's when the mineral within that sandstone formed. That's a critical difference. The mineral formed in some earlier preexisting rock which weathered to form sediments that were then transported away, accumulated in some environment, and eventually were buried and lithified to form our sandstone. When sedimentary rocks form, the minerals within them are not heated enough to reset the radioactive clocks within their constituent minerals.

How then can we talk about the age of sedimentary rocks in the Hudson Valley, or anywhere else for that matter? To answer this question, we have to understand that geologists often talk about two different ways of indicating a rock's age. One way, which we've already discussed, is called the rock's absolute age. It's how old it is as determined by a radiometric dating technique. An igneous rock in the Hudson Highlands, for example, might be given an age of 1,012 ± 4 million years (in other words, we have a high degree of confidence that the rock is between 1,008 and 1,016 million years old). The plus-or-minus age range simply reflects uncertainty due to the limitations of the instruments and techniques used in dating the rock. When the absolute age is not known, geologists can talk about the relative age of the rock. The relative age is simply how old the rock is compared to another rock. The choices are simple; it's either younger or older.

Working out the relative ages of sedimentary rocks is quite easy in principle and only requires a careful observation of their geometric relationships. Nicolas Steno (1638-1686), a brilliant polymath, worked out some of the basic principles of relative age determination we still use today. Steno was an interesting person. Born into a wealthy Lutheran family in Denmark, he became an expert in anatomy and became famous for his skills in dissection as he travelled around Europe. He eventually settled in Florence, in the Tuscany region of Italy, where he became a physician for the Grand Duke Ferdinando II de'Medici. When a large shark was caught off the nearby coast, it was sent to Steno to dissect and he noticed a striking resemblance between the living shark's teeth and objects called *glossopetrae* or "tongue stones." Steno showed how these *glossopetrae* were fossilized shark teeth – a radical idea at the time when most people believed that fossils were objects that "grew" in the rock and only incidentally resembled once-living organisms.

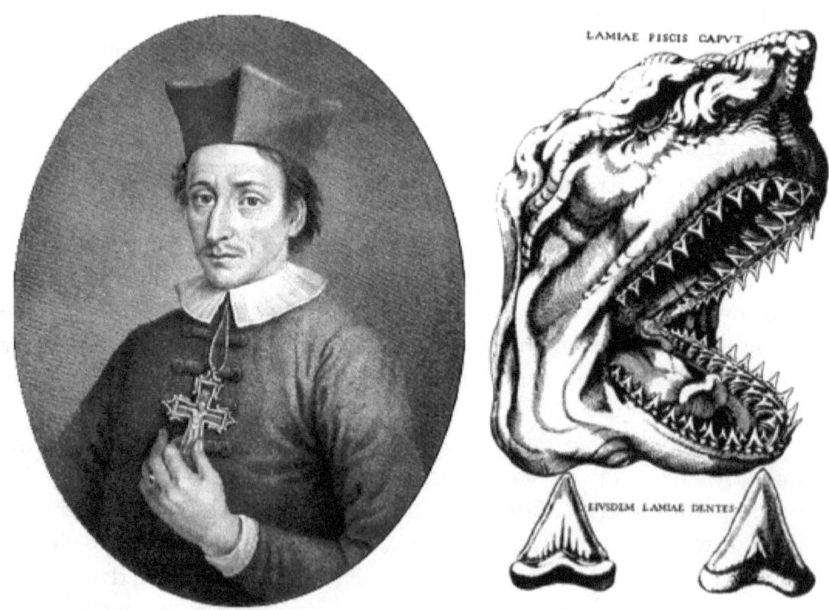

Nicholas Steno and his shark teeth illustration

Once introduced to fossils, Steno began to look at how solid objects come to be entombed within rock. This led to a study of crystal growth, for which he's also famous, and the field observations of the layering of rock strata in the area around Tuscany. His findings were presented in 1669 in a book with the unwieldy title *De solido intra solidum naturaliter contento dissertationis prodromus* (Preliminary discourse to a dissertation on a solid body naturally contained within a solid) usually referred to as simply the *Prodromus*. While Steno made numerous important contributions to science, he also felt a religious calling and left it all behind when he converted to Catholicism and was eventually consecrated as a bishop in 1677. Unfortunately, he died of a stomach ailment at a relatively young age.

What distinguished Steno from many of his contemporaries was his unwillingness to simply rely on the writings of ancient authorities, such as Aristotle, but instead to investigate everything first-hand with an open mind. The integration of scientific inquiry with his faith might best be expressed by a journal entry which read: "One sins against the majesty of God by being unwilling to look into nature's own works and contending oneself with reading the works of others." The relative age principles that Steno worked out in the mid-1600s, and which were refined by later researchers, are still used by modern geologists when examining rocks in the field. Today, these principles are called original horizontality, superposition, lateral continuity, and cross-cutting relationships. While the terms may appear intimidating at first glance, the concepts are simple, common-sense observations.

The principle of original horizontality states that sedimentary rocks were originally formed in horizontal layers called strata. This makes sense since many sedimentary rocks form by the accumulation of sediments in places like seafloors, lake beds, swamps, etc. The sediments accumulated in such places are going to be essentially horizontal. When we drive around the Hudson Valley, however, we see sedimentary rock layers in all different orientations – some are even vertically bedded. This tells us that something later happened to these rocks to alter them from their original orientation. We'll discuss how this happened in later chapters.

The principle of superposition states that when there is a sequence of layered sedimentary rocks, the oldest rocks will be at the bottom of the sequence and the youngest rocks will be at the top. It's like building a layer cake. You place layers from bottom to top as you assemble the cake. Extreme folding of these layers can, in some circumstances, completely flip rock strata upside-down and large faults can sometimes thrust older rocks on top of younger rocks but these situations are usually obvious from an examination of the surrounding area.

The principle of lateral continuity simply holds that rock layers extend laterally for some distance. This makes sense because sedimentary strata represent a particular environment. Limestone, for example, typically forms in seawater. Since the sea would have covered a large area, the limestone would extend over a large area as well. There are, however, some important caveats to this simple picture. As you move closer to the shoreline, the influx of sediment from land may gradually change that limestone into a sandstone. As you move into deeper water, limestone can no longer form and fine clays settling out of the seawater may form a shale. So, while rock strata may extend for some distance laterally, there are limits depending upon the environment in which they formed.

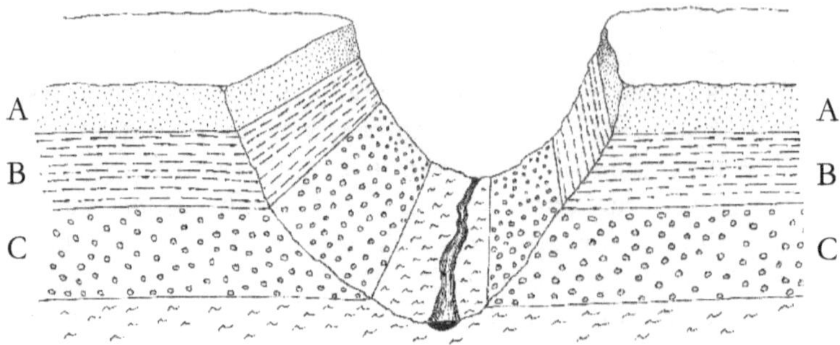

Cross section of a river valley with three different sedimentary rock units — A, B, and C. Superposition tells us layer C is older than layer B and both are older than layer A. We can correlate (match up) the rock units across the eroded canyon even though they are not physically in contact any more.

A practical application of the principle of lateral continuity is in correlating rock strata from one location to another. The White Cliffs of Dover on the southeastern coast of England are famous for their thick layers of chalk which was formed from vast numbers of marine phytoplankton microfossils. One also sees identical chalk deposits on the other side of the English Channel in France. It certainly makes sense to correlate these two rock units by imagining that they once comprised a single, laterally-continuous layer of chalk. This chalk once formed on the floor of an ancient sea with part of the layer having been removed by subsequent erosion to form the English Channel.

The final principle of cross-cutting relationships holds that when something cuts through a rock, the rock was already there (in other words, the rock is older than the structure cutting through it). In the previously discussed Hudson Highlands, there are a number of structures called dikes present in some of the rock units. A dike forms when a preexisting rock fractures and molten rock, or magma, is intruded into the fracture where it hardens into a narrow, tabular body. Dikes are obviously younger than the rock they intrude. We can make a similar argument for the younger age of faults which cut across sedimentary rock strata.

Steno's simple, yet profoundly powerful principles of relative age determination are why he's generally known as the "father of stratigraphy" (stratigraphy is the study of layered sedimentary rocks). Using these conceptual tools, geologists in Europe were then able to work out, over many decades of study, a relative-age time scale for rocks. We refer to this today as the geologic time scale.

The geologic time scale divides all of Earth history into manageable chunks of time. The largest of these are called eons and we've already mentioned three of them by name – the Hadean, Archean, and Proterozoic. Before these eons were named, this huge span of time (88% of the Earth's history!) was called the Precambrian since it referred to everything older than Cambrian Period rocks. Early geologists lumped these rocks together because they were primarily igneous and metamorphic rocks and very difficult to study before the advent of radiometric dating and modern techniques of geochemical analysis.

Around half a billion years ago, fossils of marine invertebrates rather abruptly appear in the stratigraphic record. This most-recent eon of geologic time, continuing to the present day, is thus called the Phanerozoic which is derived from the Greek for "evident life." The fossil record during the Phanerozoic Eon records a couple of catastrophic mass extinction events conveniently allowing us to divide it up into three eras – the Paleozoic ("ancient life"), Mesozoic ("middle life"), and Cenozoic ("recent life"). The end of the Paleozoic Era is marked by a still-mysterious event (or events?) which wiped out virtually all life on Earth. Some estimates for this extinction event claim that over 90% of all species living at this time disappeared forever. Later, the end of the Mesozoic Era is marked by the more famous, but less extreme, extinction event that wiped out the dinosaurs.

Close examination of sedimentary rocks and fossils they contained allowed these eras to be subdivided into periods, periods into smaller epochs, and epochs into yet smaller stages (not shown). The names of the periods, epochs, and stages have different derivations reflecting the fact that they were proposed by many different geologists studying rocks in various parts of Europe over a century of time. The Devonian Period, for example, was named after studying red sandstones in the Devonshire region of England. Jurassic comes from rocks in the Jura Mountains of Switzerland and Permian refers to the Perm area at the foot of the Ural Mountains in Russia.

Based upon relative age determinations and the correlation of rock units across wide-spread areas, geologists assigned them into their appropriate place on the geologic time scale. This was all done before radioactivity was discovered so geologists still had no idea how old these

Steven H. Schimmrich

Geologic Time Scale

Eons	Eras	Periods		Age (Ma)
Phanerozoic	Cenozoic	Quaternary		0 - 2.6
		Tertiary	Neogene	2.6 - 23
			Paleogene	23 - 66
	Mesozoic	Cretaceous		66 - 145
		Jurassic		145 - 201
		Triassic		201 - 252
	Paleozoic	Permian		252 - 299
		Carboniferous	Pennsylvanian	299 - 323
			Mississippian	323 - 359
		Devonian		359 - 419
		Silurian		419 - 444
		Ordovician		444 - 485
		Cambrian		485 - 541
Proterozoic		The Precambrian		541 - 2500
Archean				2500 - 3850
Hadean				3850 - 4540

Lengths of Eras, Periods, and Epochs to scale.

rocks were in an absolute sense. Absolute ages weren't determined for the divisions of the geologic time scale until the mid-20th century, and are still being refined today.

This discussion is important because an essential part of the education of a geology student is to become familiar with the geologic time scale. Geologists don't generally talk about 380 million-year-old fossil coral or a 520 million-year-old limestone, they instead discuss Devonian corals and Cambrian limestones. From this point forward, we will do the same as we return to our unfolding story and pick it up in the latter part of the Proterozoic Eon. Keep the Geologic Time Scale handy!

As we left the previous chapter, the Grenville Mountains were eroding away and Rodinia had begun to break apart. The breakup of Rodinia was a long and complicated affair. It stretched over a few hundred million years and many of the details are still not well understood. One of the problems with trying to study geological events that occurred in the Precambrian is that these rocks are often covered by many kilometers of younger sedimentary rocks and are therefore inaccessible except through a few, widely-scattered drill holes. Precambrian rocks exist in Ulster County, where I reside, but they're not exposed anywhere at the surface and are instead found in the deep basement over 3 kilometers below my feet.

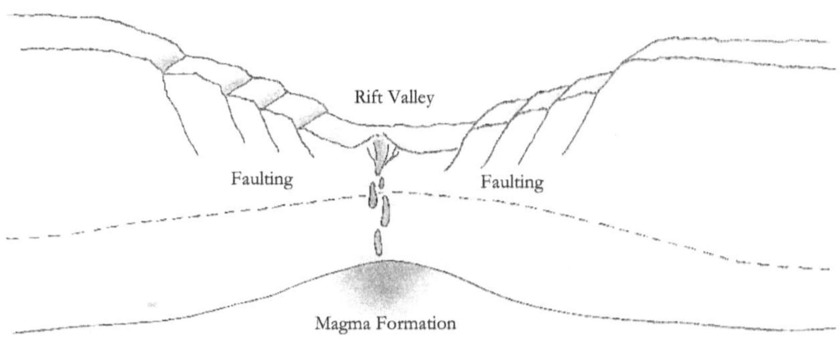

Continental rifting

Supercontinents like Rodinia are inherently unstable since their thick continental crust acts like a thermal blanket over the mantle. Heat will build up beneath the lithosphere over time and eventually initiate rifting to break the supercontinent up into separate tectonic plates. Rifting of continental crust occurs because heat flow from beneath the lithosphere, often in the form of a mantle plume, will initiate uplift and doming of the crust. As tensional forces increase in the lithospheric plate, brittle faults will begin to form in the upper crust and the lower crust will thin due to ductile deformation (flowing of the solid material). Eventually, a rift valley opens up on the surface. Rift valleys often contain volcanoes due to the high heat flow and thin crust at the center of the rift. Deep lakes can also form in rift valleys as well – look at Lake Baikal in Siberia or Lakes Tanganyika or Malawi in Africa.

When a supercontinent starts to pull apart, numerous rift valleys will form parallel to each other and perpendicular to the direction of tensional stress. Some of these valleys will eventually stop rifting ("failed rifts") while others will link up and completely fracture the continent. When the continent fractures, there is a need to create new crust if it continues to pull apart. When the lithosphere is rifted, pressure is reduced on the underlying material in the mantle. Reducing this pressure acts to lower the melting point and the mantle material begins to partially melt. Magma derived from this partial melting then rises through fractures in the existing crust and solidifies to form the igneous rock basalt. Basalt is the rock that comprises all of the world's seafloor crust and this is the process by which new ocean basins are created. Seafloor crust is much thinner and denser than continental crust and forms a low area on Earth quickly filling with seawater. Now, instead of a rift valley, we have a mid-ocean ridge on an evolving seafloor.

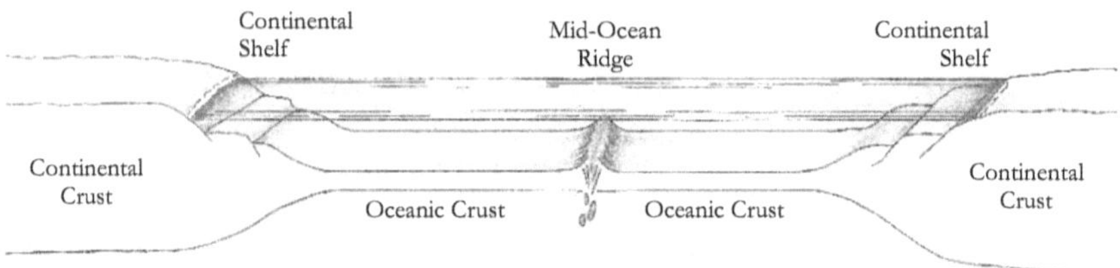

Mid-ocean ridge and seafloor spreading

Rodinia broke apart in stages with different pieces breaking off at different times over hundreds of millions of years. Faulting, rift valley formation, and rift volcanism all left their marks in the geologic record. By 600 million years ago, however, Laurentia (proto-North America) had essentially broken free but was deep in the Southern Hemisphere close to the present-day south pole.

This period of time, a little over 600 million years ago, was extremely important in the history of life on Earth. It's around this time that multicellular life, or metazoans, first appeared in the Earth's oceans. These early animals are sometimes called the Ediacara fauna after a fossil locality in southern Australia but a more general term is Vendian fauna (the Vendian is a division of geologic time in the Proterozoic Eon). The

Vendian fauna are an interesting group of fossils – while some are clearly ancient representatives of modern animals, others are much more enigmatic and harder to classify. Unfortunately, being soft bodied with no hard parts means these fossils are rare and difficult to find so our knowledge of them is somewhat limited. They are, however, the ancestors of all animals on Earth – including us.

After almost three billion years of only single-celled life (think about that for a minute!) and a billion years after the development of a cellular nucleus, why did multicellular life suddenly appear on the scene? If you remember from a previous discussion on the origin of eukaryotes, organisms with a nucleus arose around the time of the oxygen crisis. This is because environmental changes often initiate evolutionary changes through natural selection – the concept of "survival of the fittest." In the case of eukaryotes, the driving force was likely the increasing amounts of free molecular oxygen in the environment starting around two billion years ago. In the case of multicellular metazoans, another driving force may have been the mother of all ice ages – an event called Snowball Earth.

Throughout its history, the Earth has experienced a number of glacial ice ages – times when global temperatures were low enough to allow the growth of massive ice sheets on high-latitude continents. The Earth was probably too warm to support glaciations prior to 3 billion years ago. The earliest known glaciation dates back to 2.9 billion years ago and is recorded in what's now South Africa. Half a billion years later, an extensive glaciation event called the Huronian was recorded in Canada around 2.4 – 2.2 billion years ago. Then another long gap of over a billion years followed by a series of extremely intense glaciations between 850 – 630 million years ago. This period of time in the Proterozoic Eon is sometimes referred to as the Cryogenian Period (cryos is Greek for ice) and these extreme glaciations may have resulted in conditions known as Snowball Earth. During a Snowball Earth event temperatures, even at the equator, may have been below freezing resulting in a completely ice-covered Earth.

The term "Snowball Earth" was originally proposed in 1989 but was popularized a decade later by the work of Harvard geologist Paul Hoffman. Evidence for this event includes glacial deposits preserved in the geologic record on continents interpreted to be in equatorial areas at the time. The arguments Hoffman proposed for the Snowball Earth scenario are somewhat technical and will not be discussed here but basically involve the cycling of carbon dioxide between the atmosphere, oceans, and carbonate rocks (limestone). In between glaciation events ("ice-house" conditions), the Earth may have significantly warmed due to an excess of carbon dioxide – a greenhouse gas which traps heat in the atmosphere. These times of higher global temperatures are called "hot-house" conditions. While the Snowball Earth idea is still somewhat controversial, the planet may not have completely frozen over, it is clear that Earth did experience some severe climate changes during this time. And, as we've seen, metazoans appear in the fossil record shortly after this biologically stressful event.

While the Snowball Earth event is not recorded in New York State, it's important to realize it did affect our area in the sense that the crust of New York did exist during that time. Rocks in New York State record the assembly of Rodinia, the formation of the Grenville Mountains, and some faulting and mineralization occurring after the mountains formed. Then our uplifted, mountainous area entered a long period of erosion; almost half a billion years' worth of erosion – enough time to completely erode away 20 – 25 kilometers (over 12 miles) of continental crust causing the disappearance of the once mighty Grenville Mountains. Periods of geologic time, when rock is not forming, leave gaps in the rock record called unconformities by geologists. The Snowball earth event occurred during this aforementioned gap.

The Earth does not record a complete sequence of rocks from its formation to the present day. It's actually like a faulty tape recorder constantly cutting in and out and only recording snippets of the story while leaving significant gaps. Rocks form or erode away based on conditions present at that particular locality, and those conditions are constantly changing over time. The good news is that even though certain events may not be recorded here in New York, they are recorded elsewhere in the world. Geologists, by studying rocks in many different locations, are then able to splice together all of these faulty tape recordings into a single coherent story.

We are now in the Cambrian Period of geologic time. The supercontinent of Rodinia has broken apart with pieces drifting off in different directions. Later, in the Ordovician Period, dramatic events will occur in the area that's now known as the Hudson Valley but for now it's quiet. When I say "for now," I'm speaking of a time period of over 50 million years. New York has gradually become submerged with slow and gradual accumulation of sediments on what's known as a passive continental margin. A passive margin is where the edge of a continent is not a plate boundary. If a plate boundary is present, it's called an active margin. Today, the U.S. has both active and passive margins. The East Coast of the U.S. is a passive margin. A continental shelf extends outward for hundreds of miles with relatively shallow, biologically productive waters. The nearest plate boundary is out in the center of the Atlantic Ocean where the Mid-Atlantic Ridge is spewing out new basaltic seafloor crust. The West Coast of the U.S. is an active margin and we'll save that discussion for the next chapter.

Reconstruction of Laurentia (Proto-North America) during the Middle Cambrian Period (~510 Ma). Most of this continent was south of the equator. The approximate location of modern-day New York State is shown. Note that the present-day Hudson Valley was on the southern continental shelf of Laurentia.

Back to the Cambrian Period. The rifting of Rodinia formed a passive margin along what we now call the Hudson Valley and was characterized by a shallow continental shelf where sediments slowly accumulated. Unlike the present continental shelf off the northeastern U.S. coastline, the sediments which accumulated were not primarily quartz sands but rather carbonate mud. Today, this type of material is accumulating off the coast of Florida. Why did this material form in the Cambrian Period in what's now eastern New York? Between the time of Snowball Earth, when Laurentia was deep in the Southern Hemisphere, to the Cambrian Period some 100 million years later, Laurentia drifted northward toward the equator and the area that's now the Hudson Valley was then firmly in the subtropical climate zone of the Southern Hemisphere. The continent was rotated by about 90°, with the equator running from what's now Canada into Mexico, and our area was in a relatively hot and arid climatic zone facing an ocean to the south.

It would be hard to overstate the importance of the Cambrian Period in the history of life on Earth. The Cambrian was originally defined as the first period of geologic time in the Paleozoic Era because its lower boundary is so significant. The beginning of this Period is characterized by an event known as the Cambrian Explosion referring to the rather sudden appearance of marine invertebrates in the fossil record. Recall that throughout the Precambrian, life was primarily unicellular. Except for the sediment-trapping cyanobacteria forming stromatolite mounds, bacteria are rarely preserved in the fossil record and are difficult to find given their small sizes. Multicellular life appeared late in the Precambrian, recall the Vendian Fauna discussed earlier, but these organisms had no hard parts and soft squishy things are also rarely preserved in the stratigraphic record.

With the Cambrian explosion, fossils suddenly appear in rocks because some organisms developed hard parts. Hard parts, like shells, preserve much more easily than soft tissue. We are now entering the Phanerozoic Eon, named for all of the "evident life" in the fossil record. The big question for paleontologists is why hard parts developed. Or, more specifically, why did they develop at this particular point in time? While the ocean chemistry certainly played a role in the ability of marine invertebrates to pull ions out of seawater and secrete shells, it's also likely that organisms now needed them for protection. In other words, predation developed. Some animals moved beyond simple filter feeding and began to prey upon other organisms. In the fossil record, we will now see evidence of an evolutionary "arms race" that continues to the present day as animals generally evolve to become faster, fiercer, and better protected.

One of the more significant organisms to develop in the Cambrian, at least from our anthropocentric point of view, was actually still soft-bodied. These organisms were free-swimming worm-like animals that possessed a very significant adaptation, a flexible rod running along their backs called a notochord. This adaptation was significant enough that paleontologists assigned these animals to a new phylum – the Chordata. Chordates are important because they eventually gave rise to vertebrates like fish, amphibians, reptiles, birds, and mammals. These organisms are, in a very real sense, our direct ancestors. Since they're soft-bodied, Cambrian-age chordates are not commonly found in the fossil record but they have been discovered in such widely-separated places as the Northern Rockies in British Columbia (Burgess Shale), the Yunnan Province of southern China (Chengjiang), and the northern coast of Greenland (Sirius Passet).

It's important to keep in mind that all life was still in the oceans during the Cambrian Period. The continents were covered with regolith – rock and mineral fragments broken apart by physical and chemical weathering. Running water carried these weathered sediments out into the sea. While some of the quartz sand mixed in with the carbonate mud on the continental shelf seafloor here, the lighter, finer clays particles were generally carried further out into the ocean before being deposited.

Carbonate mud makes limestone and the marine invertebrates living on the seafloor became fossils entombed within the limestone. In most of the Hudson Valley, this limestone is hidden, buried under younger sedimentary rocks. In other areas, east of the Hudson River and near New York City, this limestone has also been metamorphosed into marble during a later mountain building event we'll discuss in the next chapter.

So what is limestone? It's been mentioned numerous times already but we haven't really discussed it in much detail yet. Limestone is a sedimentary rock primarily composed of a single, very common mineral called calcite. Calcite is formed from the bonding of calcium (Ca^{2+}) and carbonate (CO_3^{2-}) ions and is denoted by the formula $CaCO_3$. Since it's composed of calcium carbonate, limestone is referred to as a carbonate rock.

Limestones form in a variety of environments through a number of different processes. They commonly form in shallow, warm ocean water but can also be found in some freshwater lakes and precipitate from groundwater in caves and springs. The type of limestone we have in the Hudson Valley, however, formed in a marine environment on the seafloor. Limestone formation is strongly tied to the Earth's atmospheric composition and climate. Much of the carbon dioxide (CO_2) in the Earth's earliest atmosphere is believed to have been dissolved into seawater and then removed from the oceans via limestone formation through a series of chemical reactions. This process did not occur on Venus and Mars which is why their atmospheres are still over 95% CO_2. Limestones are also believed to have formed at much higher rates at various times in the geologic past when the Earth was warmer than it is today. A warmer Earth would have no polar ice caps, higher sea levels with huge areas of the continents flooded by shallow seawater, and a less stratified ocean. All of these conditions would have favored limestone formation directly from seawater.

In addition to the inorganic formation of limestone, it may also form from the hard parts of biological organisms. Many marine invertebrates, such as brachiopods, mollusks, and coral, build skeletal hard parts from the minerals calcite or aragonite. Aragonite has the exact same formula as calcite ($CaCO3$) but a slightly different crystalline structure. When these organisms die, most of their hard parts will eventually break down into a fine-grained carbonate mud on the seafloor. Larger fragments of the skeletal hard parts mixed into this carbonate mud may eventually be preserved as fossils within the limestone.

Limestone is usually easy for trained geologists to recognize in the field but an infallible test is to place a small drop of hydrochloric acid (HCl) on the suspected limestone. The acid will break apart the carbonate ion (CO_3) releasing bubbles of carbon dioxide gas (CO2) resulting in a fizzy froth on the surface of the rock. For this reason, field geologists often carry around small dropper bottles of HCl. If you'd like to prepare your own, muriatic acid is available at most hardware stores for cleaning scale (calcium carbonate) from plumbing pipes, reducing pH in swimming pools, and etching concrete. Muriatic acid is strong HCl, typically a 30-35% concentration. Much less of a concentration is needed to test limestones, a rough rule of thumb is that 1 part muriatic acid to 9 parts water will result in an adequate strength for geological testing. Always mix the acid INTO the water and do it slowly with eye protection and adequate ventilation! Acid fumes can sear nasal and lung tissue and splashes can burn skin and severely damage your eyes. Follow all safety instructions on the muriatic acid container. Alternatively, a friendly geology or chemistry professor might be willing to supply you with a small amount for testing purposes (ask for a 1 molar solution). When you obtain your HCl acid solution, it can be tested on a piece of concrete (which is made from limestone).

There are basically two types of sedimentary rock from the Cambrian Period exposed in the Hudson Valley Region, and neither are especially widespread at the surface. The oldest is called the Poughquag Quartzite (Poughquag is a town in southern Dutchess County). Technically, quartzite is a metamorphic rock, something we'll discuss in more detail in Chapter 5, but we're going to pretend for now that it's a sandstone since that's how it originated back in the Cambrian. The Poughquag is a very pure quartz sandstone and this rock unit appears to represent beach and shallow marine deposits from the earliest sea encroaching upon the older, highly eroded, Precambrian rocks we discussed in the previous chapter. We can infer this since small grains of the mineral zircon ($ZrSiO_4$) have been found in this sandstone with radiometric dates indicating they eroded from Grenville-age rocks. Imagine an ancient sea beginning to lap onto the flat shoreline area of a lifeless continent. The wave action washes and winnows the sediments as pure white sand collects on an empty, desolate beach.

Remember the faulty tape recorder analogy? When sedimentary rocks grade smoothly from one type to another in a vertical sequence of strata, geologists say they are conformable. In the example above, however, there's a significant gap in age between the Grenville rock (about 1 billion years old) and overlying the Cambrian-aged Poughquag sandstone (about 540 million years old). Geologists refer to contacts such as the one we see at the base of the sandstone as an unconformity (in other words, the contact is not conformable). There are different types of unconformities (we'll see another soon) but when a sedimentary rock rests directly on an igneous or metamorphic rock (remember that the Grenville rocks are primarily metamorphic gneisses), we call it a nonconformity.

Overlying the Poughquag sandstone is a thick sequence of carbonate rocks called the Wappinger Group. These rocks are also somewhat metamorphosed but we'll similarly discuss them as if they were still sedimentary. There are several recognizable subdivisions of strata within the Wappinger Group, but we're not going to worry about distinguishing them since they all formed in roughly the same type of environment – the continental shelf of Laurentia in relatively shallow marine water. In contrast to carbonate rocks formed later in the Hudson Valley, and those forming at this time elsewhere in the world, the Wappinger Group rocks have very few fossils. This is because the Wappinger Group carbonates are not limestones, they're a closely-related rock called dolostone.

Dolostones are similar to limestones but have a significant component of the mineral dolomite along with the calcite (named from its occurrence in the Dolomite Alps of northern Italy). Dolomite has the formula $CaMg(CO_3)_2$ and differs from calcite in having the addition of magnesium in the formula (it's a calcium-magnesium carbonate). In the natural world, carbonate rocks will range from pure limestones to dolomitic (or magnesiun) limestones to dolostones. The addition of magnesium makes dolostone less soluble than limestones and often results in a slightly harder rock. While experienced geologists can often distinguish dolostone from limestone in the field, the easiest way is to use hydrochloric acid. While limestone effervesces readily with HCl, dolostone will only react weakly with the acid when it's tested on powdered material (powdered rock has more surface area for reactions to occur). Think of the magnesium as holding onto the carbonate ion more strongly than does calcium, which makes it harder for the acid to break it down into carbon dioxide gas.

Laboratory studies have shown it's very difficult to precipitate dolomite directly from seawater. The problem with this is there are a lot of marine dolostones in the geologic record. How did they get there? It's believed that virtually all dolostone formed from the alteration of original limestone. This process is called dolomitization and it can occur in the presence of saline, Mg-rich brines circulating in the subsurface groundwater system. These brines can form in hot, arid conditions where seawater evaporates in areas of restricted circulation along coastlines. Dolostone is forming today in this type of coastal environment, called sabkhas, along the Arabian Peninsula in the Middle East. Our part of Laurentia was apparently in a similar subtropical climate zone during the Cambrian Period.

Unfortunately, there are not a lot of places to easily view Cambrian-age rocks in the Hudson Valley. They're present, but typically buried under younger rocks. One of the best places to see them is along Interstate 84 between the New York State Thruway (Interstate 87) and the Taconic State Parkway. As you drive east on Interstate 84, you will pass a large outcrop of white Wappinger Group carbonates near Route 9W (Exit 10). Continuing across the Hudson, look south to see Storm King Mountain on the west side of the River and the start of the Hudson Highlands where the river narrows as it flows into the harder and more resistant Grenville-age metamorphic rocks. After passing Route 9 (Exit 13), keep an eye out for Wappinger Group outcrops on either side of the road (it's best to allow someone else to drive as geologists can be notoriously poor motorists when looking at outcrops!).

SUNY Ulster students examining Wappinger Group outcrop on Lime Kiln Road off I-84 in Dutchess County

One place to get your face next to the rock may be found off Lime Kiln Road (Exit 15). Just north of the Interstate, on the west side of the road, is a nice outcrop of dolostone that can safely be examined (don't pull off to look at rocks on Route 84 unless you'd like to have a chat with a State Trooper). As we can infer from the name of this road, Wappinger Group carbonates were once mined in this area for use as mortar and perhaps also as an agricultural lime since soils in the Hudson Highlands just south of here are somewhat acidic. Heating carbonates in a kiln will drive off carbon dioxide gas and the resultant rock can be crushed into a powdered lime and spread on fields to reduce soil acidity.

One thing you will probably not find in these outcrops is fossils. As limestones are altered into dolostones, fossils are often destroyed by the growth of dolomite crystals within the original rock and dolostones are typically unfossiliferous. It's therefore not too surprising that the Wappinger Group doesn't have a lot of preserved organisms. Only a few trilobites and stromatolites have been found in widely scattered areas. Stromatolites we've already discussed since they've been around for almost three billion years but trilobites are new and worthy of a short digression since they're such an important Cambrian fossil.

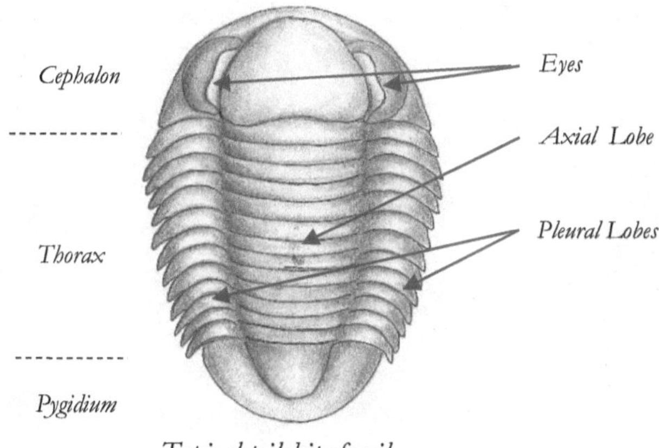

Typical trilobite fossil

Trilobites were a rich and diverse group of marine invertebrates which first appeared in the early Cambrian Period and became extinct at the end of the Paleozoic Era some 290 million years later. Trilobites belonged to the phylum Arthropoda, the largest in the animal kingdom (it's the phylum that contains all of the insects as well as a number of marine invertebrates). Arthropods are characterized by a segmented body, jointed appendages (arthropod literally means "joint foot" from the Greek), and an exoskeleton which is periodically shed during molting. While there were thousands of different species of trilobites living and dying throughout the Paleozoic Era, they all shared a number of important characteristics making them immediately recognizable as fossils.

The name trilobite comes from the fact that their bodies were subdivided into three lobes both laterally and axially. Laterally, they had a central axial lobe bordered on each side by a pleural lobe. Axially, they had a head (cephalon), body (thorax), and tail (pygidium). Many species also had well-developed compound eyes (although others were blind) and a wide range of ornamentation on their bodies (spines, bumps, etc.). In life, trilobites had antennae, feathery gills, and many pairs of legs that are rarely preserved.

While most trilobites are only a few centimeters in length (1-3 inches), they can range from a millimeter to an impressive 72 centimeters (2.4 feet). Different trilobites had differing modes of life with some swimming and eating plankton, some crawling on the seafloor as scavengers, and some burrowing in the sediment eating detritus. Because of their diversity and abundance in the Paleozoic Era, trilobite species have been used for a number of important studies of biological evolution.

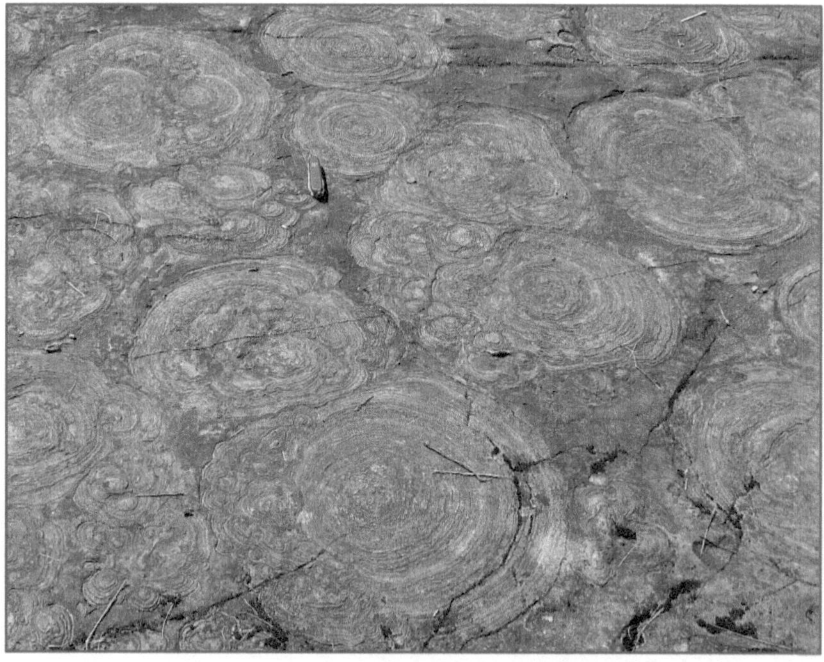

Lester Park Stromatolites west of Saratoga Springs

While Cambrian trilobites are few and far between in the dolostones of the Hudson Valley, stromatolites are easily seen if you want to travel a bit further afield. A few miles west of Saratoga Springs is Lester Park, a small roadside property owned by the New York State Museum where one can walk on an impressive outcrop covered with Cambrian Period stromatolites (*Cryptozoon proliferum*). These fossils were first studied and named by the eminent State Paleontologist James Hall in 1883 and, in recognition of its importance in New York State geology, this outcrop has been preserved as a park for almost 100 years now. While technically outside of the Hudson Valley proper, Lester Park is well worth the trip and easily accessible.

The Lester Park stromatolites are preserved in a rock unit called the Hoyt limestone which is nearly 500 million years old. Traveling back in time to the Cambrian Period would reveal a shallow coastal setting in a hot and arid climate. As the tides washed in and out, we would have seen stony mounds on the seafloor covered with greenish slime (cyanobacteria). There were a few trilobites or marine snails crawling around between mounds, but not much else. With the evolution of marine invertebrate predators, the long success story of stromatolites is coming to an end.

While the history of the Cambrian Period in the Hudson Valley is relatively uneventful, boring even, this will soon be punctuated by a short period of terror. As we move into the Ordovician Period, a chain of active volcanic islands is relentlessly approaching the shoreline of Laurentia.

4
Volcanic Islands

By turns a pitchy cloud she rolls on high;
By turns hot embers from her entrails fly,
And flakes of mounting flames, that lick the sky.
Oft from her bowels massy rocks are thrown,
And, shiver'd by the force, come piecemeal down.
Oft liquid lakes of burning sulphur flow,
Fed from the fiery springs that boil below.

Publius Virgilius Maro, *The Æneid*

Volcanoes are among the most spectacular geologic features found on Earth. Who wouldn't be thrilled to witness, from a safe distance at least, a column of ash being blown high into the sky or hot lava flowing down the slopes of an erupting volcano? Given their potential for devastation, however, it's fortunate that we don't have a lot of active volcanoes here in the mainland United States. The most recent large eruption, at least at the time of writing, was Mount St. Helens back in May of 1980. In the geologic past, however, there were times when volcanoes were erupting much closer to what's now New York State.

As mentioned in the previous chapter, the West Coast of the United States is currently an active margin which means that the coastline is near a tectonic plate boundary. The nature of this active margin coastline changes depending on whether or not you're looking at the coast of southern and central California or the coast of northern California, Oregon, and Washington State. In southern and central California, the San Andreas Fault system marks the boundary between the North American Plate and the Pacific Plate. A sliver of California, along with a large part of the Pacific Ocean seafloor, is moving northwestward in a relentless journey as it grinds past the rest of the North American continent. Each large slip of the tectonic plates along this boundary (and other associated faults) results in damaging earthquakes for those living in California.

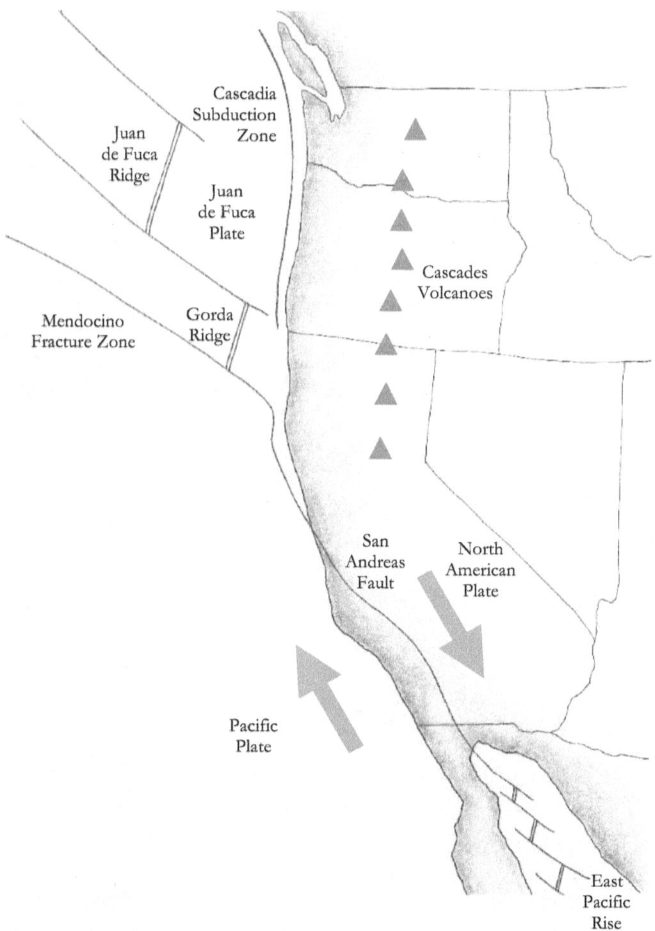

Complex tectonic plate configurations on the U.S. west coast

It's a different story in the Pacific Northwest where there is a small tectonic plate, called the Juan de Fuca Plate, just off the coast. This plate of oceanic crust is subducting beneath the North American continent.

As the subducting plate is consumed into the underlying mantle, partial melting generates magma which rises up to form the Cascades, a classic continental volcanic arc mountain chain. The Cascades are home to volcanoes like Mount Lassen, Mount Shasta, Crater Lake, Mount Hood, Mount Rainer, Mount Baker, and the now infamous Mount St. Helens. Active margins, unlike passive margins, host large earthquakes and periodic volcanic activity. Keep this in mind as we continue to move forward through the geologic history of the Hudson Valley.

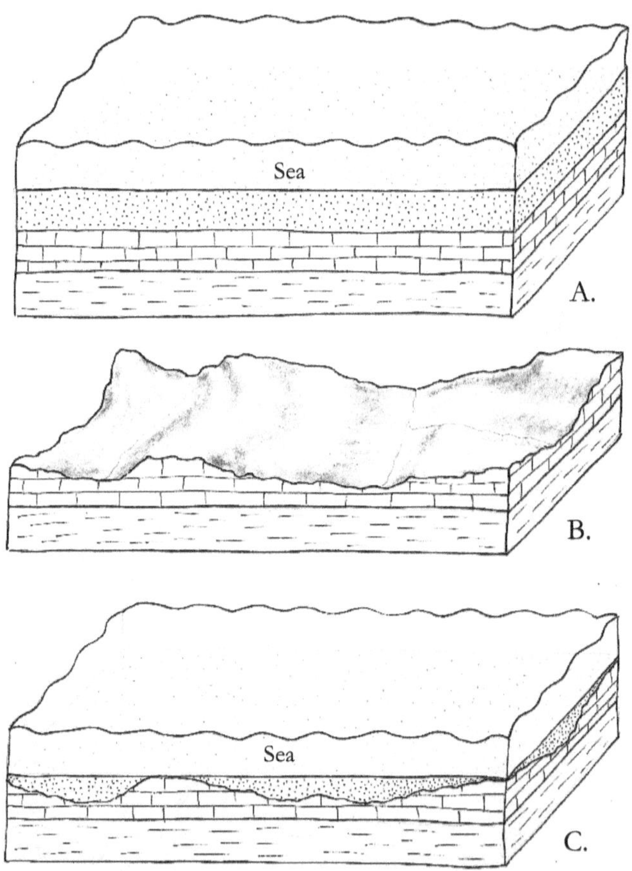

Formation of a disconformity. A: Deposition of sedimentary rock layers beneath the sea. B. Regression of the sea, exposure as dry land, and erosion of the layers. C. Transgression of seawater and deposition of new sedimentary layers on the old erosional surface.

One of the things we see in North American geology during the Paleozoic Era of geologic time are several worldwide fluctuations in sea level. When the sea level rises, we have a marine transgression (seawater advances onto the land surface). A marine regression occurs when the sea levels fall and seawater recedes off the land surface. During a transgression, marine sedimentary rocks are deposited on the continent – that's why we can find marine limestones in places like Ohio or Kansas. When sea levels drop, erosional surfaces develop on these sedimentary rocks resulting in unconformities. Earlier we mentioned that sedimentary rocks deposited on igneous or metamorphic rocks form a type of unconformity called a nonconformity. This is what formed between the sedimentary Poughquag sandstone and the metamorphic Grenville crust during the Cambrian Period. Unconformities that develop due to marine regressions, and form between two parallel layers of sedimentary rock, are called disconformities. Once again, we can invoke the faulty tape recorder analogy where sedimentary rocks faithfully record marine transgressions but nothing is recorded during a marine regression. As a matter of fact, some of the evidence is actually destroyed during a regression due to erosion removing previously-deposited rock layers.

Since we see evidence for certain marine transgressions and regressions in other parts of the world, as well as here in New York State, they're reflecting global sea level changes. Keep in mind, however, that these are not global floods – even during the highest marine transgressions, there were areas of dry land on the continents. In addition, the geologic evidence shows that these transgressions were gradual events; if you were living during these times you'd notice nothing more in your old age other than the fact that the coastlines were slightly further inland than they were when you were a child; something we may all experience soon if global warming continues as it has been in the recent past.

There are two primary causes of global sea level change – glaciation and plate tectonic activity. When the Earth's climate cools and large glaciers form on the continents during an ice age, global sea levels drop since the ice in glaciers forms from large accumulations of snow. Snow results from atmospheric moisture and this water vapor is ultimately derived from evaporating seawater. During an ice age, that water is trapped in glacial ice and doesn't return to the sea to replenish what was lost to evaporation. When global conditions are warmer than normal, there may be no glaciers on Earth and sea levels will be higher because glaciers will all melt adding significant volumes of water to the oceans.

Plate tectonic activity can affect sea level as well. When there are a lot of mid-ocean ridges on the seafloor, sea levels will be higher. Mid-ocean ridges are places where the seafloor is ripping apart due to tensile stresses from spreading tectonic plates (seafloor spreading). Magma rises to fill in the void and cools to form new basaltic seafloor. At the ridge, however, the hot, relatively-thin lithosphere bulges upward to form a seafloor mountain range. At the present-day Mid-Atlantic Ridge, where the North and South American Plates are splitting away from the Eurasian and African Plates, the crest of the ridge is, on average, 2.5 kilometers above the deep abyssal plain seafloor and over 1000 kilometers in width. Now when we consider that there are over 70,000 kilometers of mid-ocean ridges in the Earth's oceans, this adds up to a lot of rock – if we assume a roughly triangular shape for the mid-ocean ridges, we can estimate a volume of 8.75 x 107 cubic kilometers when everything's multiplied together). Since the volume of water in all the world's oceans is about 1.3 x 109 cubic kilometers, the mid-ocean ridges comprise over 6% of the total volume of the ocean basins. That's significant. When plate tectonic configurations exist such that there are many mid-ocean ridges, sea levels will be higher, and water will transgress onto the continents, than when there are fewer mid-ocean ridges on the seafloor. It's like placing a rock into a full bucket of water – the rock displaces some of the water which will then spill out over the sides of the bucket.

The first of the marine transgressions of the Paleozoic Era began in the waning days of the Precambrian and peaked in the Late Cambrian and Early Ordovician Periods. Geologists call this the Sauk transgression and shallow seas covered much of Laurentia during this time. This rise in sea level was likely due to the breakup of Rodinia, which formed many new mid-ocean ridges, with a contribution, perhaps, of meltwater from the last of the earlier Snowball Earth glaciers.

As marine water transgresses onto a continent, we typically see a distinctive sequence of quartz sandstones grading upward into carbonate rocks. We already spoke about how we see the sequence of Poughquag Quartzite overlain by Wappinger Group carbonates in the Hudson Valley, but it's important to realize that we also see this same type of sequence throughout North America. Elsewhere in New York we see Potsdam Sandstone surrounding the Adirondacks and grading upward into the Little Falls Dolostone. This dolostone is famous for containing Herkimer "diamonds," which are actually not diamonds but beautiful quartz crystals precipitated in cavities within the rock from the ancient circulation of silica-rich groundwater. In places like Ohio, we see the Cambrian Mt. Simon Sandstone grade upward into a carbonate unit called the Knox Dolostone. Even as far away as the Grand Canyon in northern Arizona, we see the Tapeats Sandstone grade upward into Bright Angel Shale and then into the Muav Limestone as the Sauk transgression swept through the western part of North America. What we see here in eastern New York did not occur in isolation, it's part of a larger worldwide sequence of events.

As we pass from the Cambrian Period into the Ordovician Period, there's initially little change in the environment of eastern New York where carbonate sediments are accumulating on a passive continental shelf. Proto-North America, or Laurentia, was rotated 90° and our area was in the subtropical zone of the Southern Hemisphere. The ocean off to our south has been called the Iapetus Ocean by geologists (Iapetus was the father of Atlas for whom our modern Atlantic Ocean is named). The only way geologists can recognize that the Cambrian/Ordovician geologic time scale boundary was passed in this area is that the assemblage of fossils found in the carbonates changes from those typically seen in Late Cambrian rocks to those seen in Early Ordovician rocks. In the Early Ordovician Period, around 475 million years ago, sea levels began to fall as plate configurations changed and the shallow seas which covered Laurentia began to drain off, leaving behind a widespread erosional surface. In the eastern U.S., geologists refer to this erosional surface as the Knox unconformity since it forms the top of the Knox Dolostone in places like Ohio but it also caps the Wappinger Group carbonates here in the Hudson Valley.

Why did sea levels drop around this time? There's evidence, from elsewhere in the world, that a large remnant of Rodinia, a Southern Hemisphere supercontinent called Gondwana, drifted across the South Pole during this time. Just as in Antarctica today, large continental glaciers developed on Gondwana which resulted in a lowering of sea levels around the globe. This event also correlates with a mass extinction event in Earth history where numerous groups of marine invertebrates suddenly disappear from the fossil record. Ocean temperatures would have cooled during this ice age and, as water levels dropped, shallow seas dried up reducing the amount of marine ecosystems available and greatly increasing competition for survival among different species. This is yet another example of environmental change driving the evolutionary change we see in the fossil record.

This is where things get complicated. We're approaching the middle of the Ordovician Period. In the Hudson Valley, the Ordovician is characterized primarily by two things. The first is the deposition of shales and sandstones in relatively deep ocean water. The next is the approach and collision of a chain of volcanic islands dramatically folding and faulting these earlier sedimentary rocks. Those two little facts, however, can be expanded upon for the rest of this chapter.

Let's back up to the Late Cambrian Period. While carbonates were being deposited near what's now the Hudson Valley, deep water shales and sandstones were being deposited further offshore (east of us today, south of Laurentia back then). As we move into the Early Ordovician Period, we start to see the appearance of these deeper water sediments in our area. What's happening?

Reconstruction of Laurentia (Proto-North America) during the Middle Ordovician Period (~470 Ma). Most of this continent was south of the equator. The location of modern-day New York State is shown. Note the chain of islands approaching from the south.

It turns out that a chain of volcanic islands had developed off the coast of Laurentia by the formation of a subduction zone out in the Iapetus Ocean. We discussed subduction earlier when talking about the formation of the Grenville Mountains when ocean crust subducted beneath continental crust. In this case, however, ocean crust is subducting beneath ocean crust forming volcanic rocks with a different composition. As with subduction beneath continental crust, subduction beneath oceanic crust results in a deep seafloor trench, large earthquakes as the plate grinds its way down, melting of the subducting slab, and a chain of volcanoes at the surface to one side of the trench. These volcanoes form islands and are called volcanic island arcs by geologists.

A modern-day example of this type of tectonic setting is the Aleutian Islands extending southwestward from Alaska. The Aleutians are volcanic islands which formed from the subduction of oceanic crust of the Pacific Plate beneath oceanic crust of the Bering Sea (part of the North American Plate). This subduction zone is marked at the surface by the deep Aleutian Trench which is located just south of the islands. In the subsurface, a dipping zone of earthquakes extends from the trench to beneath the islands marking the passage of the subducting oceanic crust as it moves downward to the north.

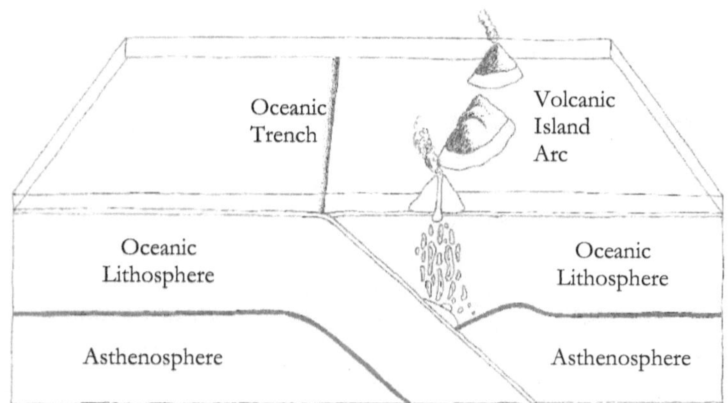

Aleutian island arc subduction

Geologists refer to the volcanic island arc which formed off the coast of Laurentia during the Ordovician as the Taconic Arc. Like modern island arcs, we can infer that the Taconic Arc experienced periodic earthquakes and large volcanic eruptions as subduction proceeded over time. We see a change from shallow water carbonates to deeper water shales and sandstones in the Hudson Valley area because water depths generally increased as the island arc trench approached our area.

As a matter of fact, as the crust flexed downward with the approach of the island arc, fractures developed in the seafloor which erupted small amounts of basaltic lava. When lava is erupted on the seafloor, it forms distinctive structures called pillows. This is due to the fact that hot lava (about 1,200° C) hitting cold seawater will almost instantly form a glassy rind and extrude as rounded masses. Video of modern seafloor eruptions show extruding lava looking almost like toothpaste being squeezed out of a tube. Ordovician pillow basalts which formed by this process have been found in New York State at a location called Stark's Knob near Schuylerville in Saratoga County.

Stark's Knob is named after Revolutionary War General John Stark and played an important strategic role in the Battle of Saratoga in 1777. Stark's Knob, and other hills in the area, form a natural bottleneck between them and the Hudson River to the east. As British General John Burgoyne attempted to retreat north back to Fort Ticonderoga after failing to dislodge the colonists on the Saratoga battlefield, the Americans "corked the bottle" by blocking the British retreat and Burgoyne had no option but to surrender – a critically important turning point in the war.

Today, Stark's Knob is owned by the New York State Museum and preserved as a small park both for its historical and geological history. A short walk from the parking area off Route 32 will bring you to an old quarry with an impressive rock face. A close examination of the quarry wall will reveal that it's composed of a pile of rounded pillow basalts. Geologists studying this outcrop have even found Late Ordovician-age marine snail fossils in small blocks of limestone incorporated into the basalts establishing its age between 460-440 million years old. This is consistent with the timing of the approach of the Taconic island arc.

As water depths increased, a number of different sedimentary rock types were being deposited on the seafloor here. The first, and simplest, is shale. Shale is the most common sedimentary rock on Earth because it forms from clay-rich mud. Clay is a generic term referring to a whole host of different minerals. What they all have in common, however, is that they tend to form by the chemical weathering of other rock-forming minerals such as the feldspars – minerals that were common in the earlier Grenville-age metamorphic rocks.

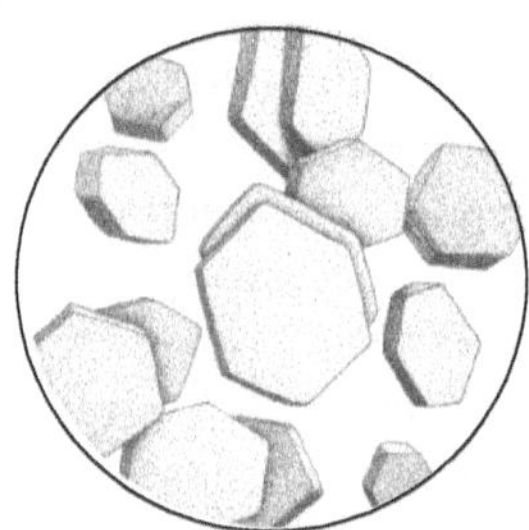

Platy nature of clay minerals

Clay minerals are also exceedingly tiny and best viewed with an electron microscope. Under magnification, clays are typically seen as flat, hexagonal plates and this platy nature of clays is what imparts fissility to shales, the property by which they tend to split into thin chips.

Since mud can accumulate in a variety of environments, it's not always easy to tell how a particular shale formed. Fortunately, fossils and other structures within the rock can provide clues to help us work out the shale's ancient environment of deposition. Ordovician shales in the Hudson Valley often contain fossils of small marine organisms known as graptolites which enables us to determine that they formed in relatively deep ocean water. Shales form in deep water because the very small and light clay minerals are the only sediments able to drift that far out into the ocean where they slowly settle through the water column to accumulate on the seafloor.

Shales can be seen throughout the Hudson Valley north of the Hudson Highlands, especially near the River. Note, when looking at these shales as you drive around the Valley, that they are usually not nicely bedded into flat, horizontal layers. While they were originally deposited this way (remember the principle of original horizontality?), the later collision of the Taconic island arc heavily folded and faulted them. An early name for these Ordovician rocks was the Hudson River Shales.

Graptolite fossils are easily missed by those not specifically looking for them in Hudson Valley shales. They appear like nothing more than small (a few millimeters in length), black pencil marks that resemble miniature hacksaw blades. Their appearance gave rise to their common name of graptolites (from the Greek words graphein, "to write" and lithos, "stone"). Graptolites existed throughout the Paleozoic Era but are very important in the correlation of Ordovician rocks in New York because of their abundance and diversity during this time.

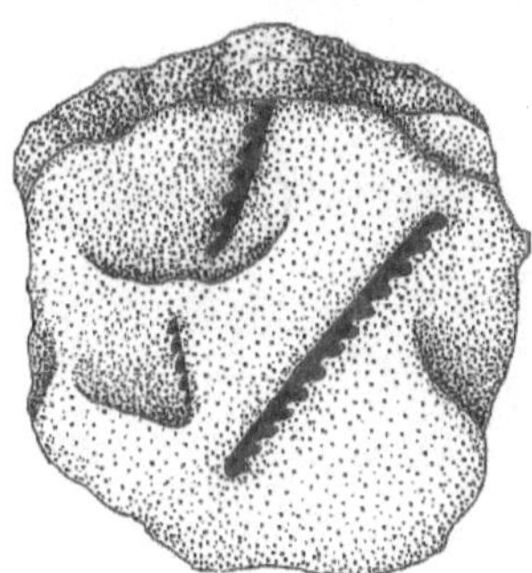

Graptolite fossils

Graptolites were interesting little animals. They're classified into the phylum Hemichordata because they had the primitive beginnings of a notochord similar to an obscure group of modern marine organisms known as pterobranchs. The animals themselves were practically microscopic and lived in colonies with hundreds of individual zooids. Those miniature hacksaw blade fossils represent structures called stipes and each of the hacksaw "teeth" was actually a tubular chamber in which an individual zooid lived. In life, graptolite stipes apparently hung from a gas filled bag of tissue and floated near the surface of the oceans where the zooids fed on plankton. This floating, or planktonic mode of life allowed graptolites to easily spread over large areas of the ancient world.

Since graptolites are today extinct (although living pterobranchs are thought to be close relatives), not a lot is known about the actual animals. They are very useful, however, as index fossils in Ordovician rocks because of their widespread range and the fact that they evolved quickly with many different species existing during this period of time. Geologists use different species of graptolites to differentiate between otherwise indistinguishable rock units. In the Hudson Valley, for example, there are Orthograptus ruedemanni, Corynoides americanus, and Diplograptus multidens graptolite zones characterizing different strata of Ordovician shale by the particular species of graptolite fossils they contain.

The occurrence of graptolites and the absence of most other marine fossils in these Hudson Valley shales indicate that the bottom waters were apparently hostile to life. This can happen in deep marine basins where the water has poor circulation and becomes stratified. The seawater near the bottom then becomes depleted in dissolved oxygen (anoxic) and marine invertebrates then have a hard time existing on the seafloor. Graptolites, because they floated in the surface waters, had no difficulties and their stipes accumulated on the seafloor because there were no bottom-dwelling scavengers to break them down before they became fossilized.

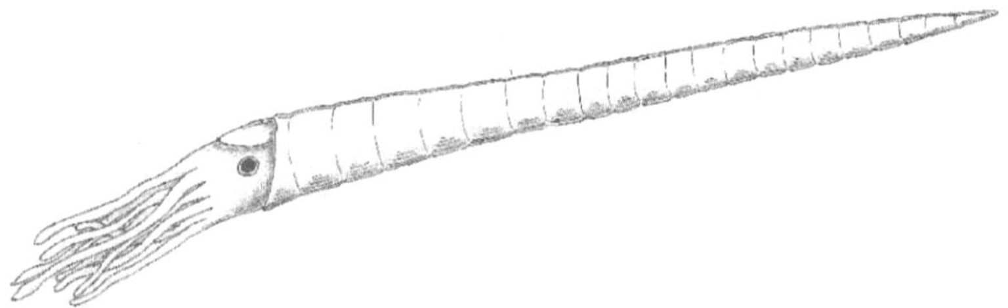

Reconstruction of an Ordovician nautiloid

While not found, to my knowledge, in Hudson Valley shales, Ordovician rocks in central New York contain fossils of large, squid-like animals called nautiloids. One well-known example is Endoceras annulatum which thrived, as a species, for several million years during the Middle Ordovician Period. These mollusks had a straight shells, some growing to several feet in length, and were only able to thrive because they swam in the sunlit, oxygenated waters closer to the surface. These animals were the top predators of the Ordovician seas.

Another type of Ordovician rock found in the Hudson Valley is sandstone. Sandstone is also a generic term in geology. It denotes rock which is composed of sand-sized grains of sediment cemented together.

The most common mineral in sandstones is quartz. Quartz, as you remember, is quite abundant in the crust of the Earth since it's found in most rocks. Quartz is also highly resistant to both physical and chemical weathering and tends to persist in the environment for a long time. Both of those characteristics lead to quartz being the dominant mineral in sediments formed by weathering of preexisting rocks. While some sandstone is composed almost entirely of quartz grains – quartz sandstone – other sandstones have varying amounts of additional minerals present. Most of the Ordovician Period sandstones in the Hudson Valley are not clean quartz sandstones, but rather what geologists call dirty or immature sandstones. These sandstones, in addition to the ubiquitous quartz, also have a fair amount of other mineral fragments together in a clay-rich matrix. These dirty sandstones are commonly known as graywackes – a word derived from German and meaning, simply, "gray rock." Graywackes typically form from submarine landslides known as turbidity currents.

In 1929, there was a large earthquake on the edge of the Grand Banks, an area of the continental shelf off the southern coast of Newfoundland. A few minutes after the earthquake, several transatlantic telegraph cables between Newfoundland and Europe began to break on the continental slope off the Grand Banks. By examining the timing of the breaks and their distance from each other down the slope, scientists were able to later determine that they were broken by a 60 mile per hour submarine landslide of sediment triggered by the earthquake. These submarine landslides, or turbidity currents, have since been filmed on the continental slope – the area where the relatively shallow continental shelf slopes downward into the deep ocean abyssal plains.

Turbidity currents occur because the accumulation of sandy sediments along the edge of the continental shelf may eventually become unstable and start to slide downslope. As the sediments move, they become entrained in water forming a murky cloud of material. Being more dense than clear water, they will quickly flow down the continental slope. Turbidity currents may be caused by earthquakes (as in the 1929 event) but seem to periodically occur in many areas without any obvious triggering events. The continental slope is, in many places, characterized by submarine canyons which were scoured by repeated turbidity flows in an area. The base of the slope, called the continental rise, is where turbidity currents spread out and deposit their sediments into wedge-shaped submarine fans.

When turbidity current sediments settle out of the water column at the base of the continental slope, the heaviest particles settle out first followed by progressively finer sediments. This results in a thick layer of sand exhibiting what we call graded bedding. If you would like to make your own graded bedding, get some small pebbles, sand, silt, and clay and toss it all into a jar of water. Shake the jar and set it down on the table until the water clears. When you examine the sediments at the bottom of the jar, you'll see coarse pebbles grading upward into fine clays. When this type of sediment lithifies, or turns into sedimentary rock, it will form a graded bed of graywacke sandstone.

Geologists have a special name for a repeating sequence of deep-water shales interbedded with graded beds of graywackes. These packages of rock are called turbidites, obviously reflecting their origin as turbidity currents. The graywacke sandstone beds in turbidites also commonly exhibit another feature which belies their origin as turbidity currents. These features are called sole marks (as in the sole of your shoe, not your eternal soul) or flute casts. Sole marks are elongate bulges on the bottom of the sandstone beds which formed due to scouring from the turbidity flows. As the currents ripped through an area of muddy seafloor, they scoured depressions which then filled with sandy sediments. The casts of the scour marks are today exposed as bulges at the base of the sandstones.

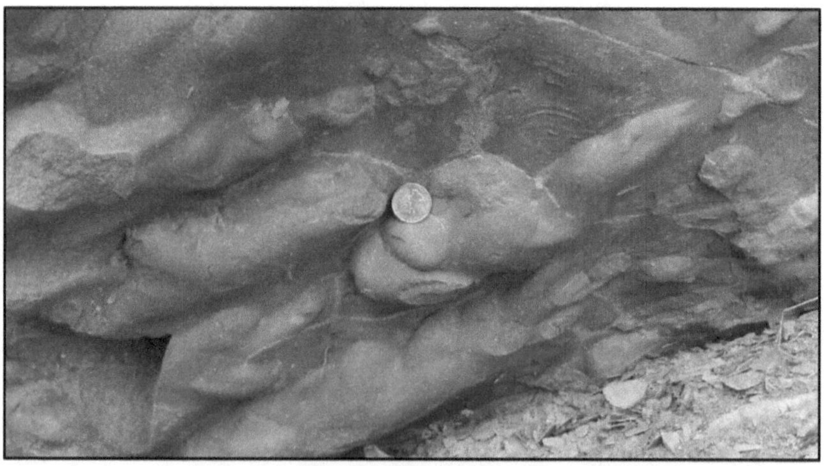

Sole marks on the bottom of a graywacke turbidite bed

Sole marks can tell us a number of useful things about the rock unit. They indicate that the rock formed from a turbidity current which tells us something about the ancient environment (a continental slope). They indicate the bottom of the bed. While this sounds obvious, it's not always easy to tell which side of a sandstone bed is the bottom when the beds have been tilted into a vertical orientation due to later mountain building.

And, finally, they indicate paleocurrent directions. Studying sole marks over a large area can tell us which general direction the turbidity currents were coming from. In the case of Ordovician graywackes in our area, they appeared to be mostly from the south (present-day east) – the vicinity of the Taconic island arc.

A great example of a turbidite sequence is found on the west side of the Hudson River along the approach to the Mid-Hudson Bridge (also named the Franklin D. Roosevelt Bridge after the president who lived in nearby Hyde Park) which connects Highland in Ulster County to Poughkeepsie in Dutchess County. Traveling from west to east on Route 44/55, you pay a toll and then drive past a spectacular outcrop of gray sandstones and shales before crossing the bridge. Since you're not allowed to stop here without special permission from the Bridge Authority, have someone else drive, obey the rather slow speed limit, and you'll be able to see that these rocks are not horizontal but tilted, in some places almost vertically.

Turbidite sequence near the Mid-Hudson Bridge approach in Highland. Thicker beds are graywacke sandstones and thinner laminated beds are shales.

Even a casual examination of these rocks reveals dozens of repeating layers – sandstone, shale, sandstone, shale, sandstone, shale seemingly ad infinitum with each layer having a different thickness. The shales are typical of those formed by slow accumulation of clays in deep ocean water and contain graptolite fossils. The sandstones, however, tell a different story. The sandstones are graywackes that formed very quickly as massive turbidity currents. You're looking at rocks that formed near the base of the Ordovician continental slope!

A nice place to see sole marks, although you're also not allowed to stop, is a bit further up the Hudson Valley on the side of the ramp from the New Baltimore Travel Plaza (mile 127) back onto the New York State Thruway heading south. As you drive along the entrance ramp, the outcrop on the right (best seen from the passenger seat) shows beautiful bulbous sole marks on the base of a tilted graywacke sandstone bed. This is essentially the same group of rocks that are exposed down by Poughkeepsie 60 miles to the south. Nearby, along Route 9W, a mile or so north of the Coxsackie Thruway entrance, is a similar outcrop where you can actually get out and look at more sole marks. This outcrop is sometimes jokingly referred to as the "dinosaur skin" outcrop due to the wrinkled appearance of the flute casts. The name came from the claims of a local high school science teacher that these surface features represented a fossil impression – he was way off, by the way, since these rocks formed more than 200 million years prior to the appearance of the earliest dinosaurs.

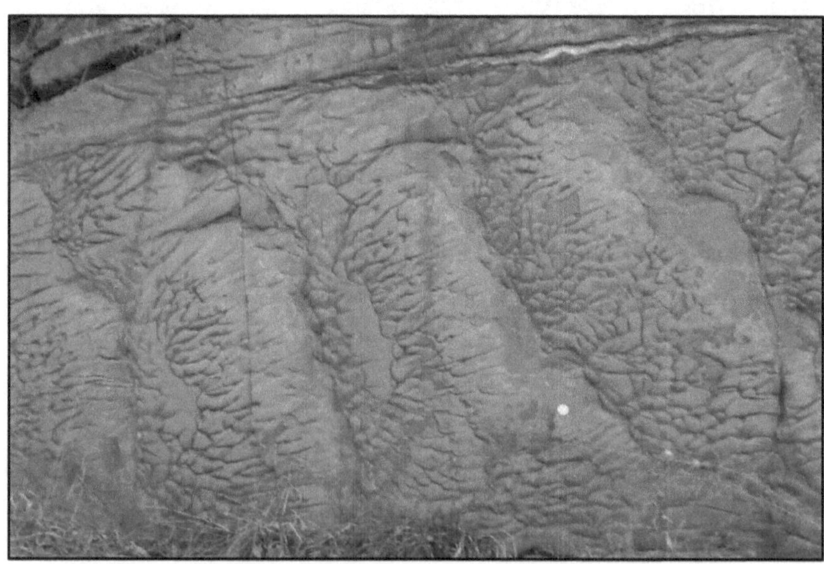

"Dinosaur skin" outcrop along Route 9W near Coxsackie. Bumps are sole marks on the underside of a tilted graywacke sandstone bed. Quarter for scale.

A final Ordovician Period rock type of great importance in the Hudson Valley is mélange. The word is from the French for a heterogeneous mixture and is quite apropos for the rock it describes. Mélanges are composed from large, generally angular blocks of rock in a finer-grained matrix of sediment. They typically form near a subduction zone trench where sediments are being scraped off of the subducting plate and blocks of oceanic rock and muddy sediments are tumbling down nearby slopes due to frequent earthquakes and volcanic activity of the nearby island arc. This type of environment is called an accretionary wedge. Stark's Knob, mentioned earlier, is actually embedded within a surrounding mélange and the numerous Ordovician mélanges in the Hudson Valley again attest to the fact that this area was characterized by the approach of a volcanic island arc. Rocks exposed in Kaal Park, on the east side of the Mid-Hudson Bridge in Poughkeepsie, also represent a mélange.

Even more compelling evidence that volcanoes were lurking nearby, however, is the presence of something else in our local rocks from the Ordovician Period. Interspersed throughout many of the Hudson Valley shales are thin layers of clay called bentonite. Bentonite is altered volcanic ash. As volcanoes in the approaching Taconic island arc periodically erupted, large amounts of ash were deposited in the surrounding seas. As this ash settled to the seafloor, it became incorporated into the bottom sediments where it chemically altered into a type of clay.

The neat thing about bentonite clay layers is that they can be radioactively dated since they still contain minerals which formed in the magma chamber of an island arc volcano. Different clay layers can also be geochemically "fingerprinted" allowing us to correlate ash layers from specific eruptions over widely separated areas. These layers give us absolute time horizons within thick sequences of sedimentary rocks which are otherwise undatable by radioactive isotopes.

One bentonite clay layer, called the Millbrig Ash, dates from 450 million years ago (the Late Ordovician Period) and is traceable from New York as far westward as the Mississippi Valley area. It even appears to correlate to an Ordovician ash layer in Sweden called the "Big Bentonite." While Scandinavia was not quite so far from North America during the Ordovician Period, it was still a good 1,000 km (600 miles) away on the other side of the Iapetus Ocean. The Millbrig Ash/Big Bentonite represents a massive volcanic eruption, one of the largest ever identified in the geologic record, emitting well over 1,000 cubic kilometers (240 cubic miles) of ash. That's a cube of ash over 10 kilometers (6 miles) on each side! For comparison, the 1980 eruption of Mount St. Helens emitted less than 3 cubic kilometers of ash and the massive 1815 eruption of Tambora in Indonesia, which altered global climate for over a year, emitted only 100-150 cubic kilometers. Dozens of Taconic age ash layers have been identified in eastern North America. Between the massive volcanic eruptions and large earthquakes, New York would often have been a terrifying place during the Ordovician!

Back to Hudson Valley rocks. In the previous chapter, we introduced the Wappinger Group of dolostones which formed during the Cambrian Period. A close examination of the Wappinger Group, however, reveals that it's not all the same. There are at least three distinctive subunits that can be traced for some distance across Dutchess County. These are referred to as the Stissing, Pine Plains, and Briarcliff Formations. They're all essentially dolostones, but can be distinguished from each other in the field based on other characteristics (the Briarcliff can have some chert in it, for example). Rock formations are the basic unit of stratigraphy and are given names based on a nearby town or geographic feature. To be called a formation, a rock unit must be recognizable in the field and extensive enough to be placed on a geologic map of the area. A number of related formations can be designated as a stratigraphic group (like the Wappinger Group). In addition, a stratigraphic formation can be subdivided into members if it has its own recognizable subunits.

If it all sounds very confusing, that's because it is. Sediments are deposited in various environments based upon the current conditions but conditions change over time. When we look at a vertical sequence of rock strata, we want to understand how and when each part of it formed. To do this, we impose our man-made classification schemes and subdivide this vertical pile of rocks into groups, formations, and members. The problem is that any two geologists may do this differently based upon what they see as important characteristics in the rock layers. In addition, some geologists are "lumpers" and some are "splitters" when it comes to applying stratigraphic classifications. A "lumper" might lump together all the associated limestones and call it a formation. A "splitter" might reasonably notice that there are subtle differences in the limestones, call it a group, and then subdivide out various formations and members. Neither is incorrect – it just depends on what your purpose is in studying the rocks. If you just need to note that all the limestones were deposited in a shallow sea which covered the area at some time, the "lumper" scenario might work fine for you. If, however, you want to study the advance and recession of that sea over time, the "splitter" scenario would be better because those subtle differences in the limestone beds are actually representing things like changes in seawater levels and wave energy reaching the bottom sediments.

Stratigraphers are forever arguing about whether or not something should be designated as a member or a formation or a group (remember, it's not what it is, it's what we choose to designate it – there is no "correct" classification). I bring this up as we discuss Ordovician rocks in the Hudson Valley because the stratigraphy of the Ordovician is confusing to say the least. While most of the Ordovician rocks in the Hudson Valley are simply shales and sandstones, geologists will bitterly argue about whether or not two identical-looking shales from different areas of the Valley belong to one formation or to another. Sometimes, they'll even disagree as to the proper names for the formations. Geologists working on these rocks over the past century have not been consistent in the names applied to different strata and revisions are common. Names like Bushkill, Snake Hill, Martinsburg, Normanskill, Austin Glen, Quassaic, Mt. Merino, and Indian River are tossed about referring to rocks that, to a layman, look pretty much the same. Fortunately, we don't have to worry about the details. For our purposes, I'm simply going to subdivide the Ordovician Period sedimentary rocks of the Hudson Valley into three main groups – the Normanskill Group, the Martinsburg Group, and the Quassaic Group. Keep in mind that there are other geologists who might reasonably disagree with my choices as being overly simplistic!

The Normanskill Group is the oldest and formed during the early part of the Middle Ordovician. It's exposed along both sides of the Hudson River essentially from the Hudson Highlands northward past Albany. It's more extensive on the east side of the River and is composed primarily of shales and sandstones. Some units are also rich in chert, a fine-grained form of quartz that archaeologists often refer to as flint. As an aside, at a place called Quarry Hill in West Athens, between Catskill and Coxsackie on the west side of the Hudson River, Native Americans mined chert from the Mount Merino Formation for knapping into arrowheads, knives, and spear points. Besides the Mount Merino, another important unit in the Normanskill Group is called the Austin Glen Formation.

The Martinsburg Group formed in the latter part of the Middle Ordovician and is exposed on the west side of the Hudson River from just north of New Paltz southward into Pennsylvania. It's bordered on the southeast by the Hudson Highlands and on the west by the Shawangunk Ridge. The Bushkill Shale is an important formation within this group.

The Quassaic Group is the youngest and appears to have formed early in the Late Ordovician. The boundary between the Middle and Late Ordovician was about 460 million years ago and we can keep that in mind as the approximate age of these rock units. The Quassaic is mostly sandstone and exposed in a narrow belt (which is actually a down-warped fold) from Kingston down to Newburgh. A line of small hills traces the aerial extent of this resistant rock unit – Hussey Hill, Shaupeneak, Illinois Mountain, and Marlboro Mountain from north to south.

There are two factors which make the stratigraphy of these rocks so complex. The first is that they're all very similar. Out of context, a chip of shale from the Normanskill Group may look identical to a chip of shale from the Martinsburg Group. It's only by studying them in context, in the field, that we can unravel their relationships. Secondly, these rocks have been heavily folded and faulted. As a matter of fact, many if not most of these rocks are not presently found where they originally formed. A large mountain-building event, which we'll discuss shortly, shoved massive slices of these rocks from east to west during the Late Ordovician Period. This folding and faulting has, in some areas, emplaced older rocks on top of younger rocks making it impossible to apply the principle of superposition. Thrust slices may also pile up within the same rock unit making it difficult to determine its original pre-thrusting thickness.

Now that we've discussed the stratigraphy of these rocks, let's look at one major reason why the stratigraphy is so difficult to unravel – their deformation by the collision of the Taconic island arc in the Late Ordovician Period.

The area that's now New York State was mostly submerged beneath ocean water during the Ordovician Period as global sea levels began to rise. Remember the Sauk marine transgression of the Cambrian Period discussed in the previous chapter? A second marine transgression, called the Tippecanoe, starts during the Middle Ordovician Period, and peaks by the Early Silurian Period. In the Midwestern United States, this transgression is marked by very pure (99% quartz) shallow marine sands called the St. Peter Sandstone which are overlain by thick sequences of fossiliferous carbonate rocks. In much of New York, clays accumulated in deeper waters to form 300 meters (1,000 feet) of shale called the Utica Shale Formation and limestones formed in shallower waters to form the Trenton Group carbonates. This sea covering New York is often called the Trenton Sea and these rocks are full of marine invertebrate fossils.

The Hudson Valley, however, was dominated by the approach of the Taconic island arc during this time. Waters were deeper in an elongate north-south anoxic basin as the ocean floor flexed downward due to the approach of the arc. This flexure caused the eruption of pillow basalts in the Stark's Knob area. Frequent large earthquakes and massive volcanic eruptions would have shaken the region sending turbidity currents and mélange blocks cascading down into the basin. This is where Normanskill Group turbidites were accumulating as the chain of volcanic islands relentlessly approached.

Finally, around 460 million years ago, these volcanic islands start to impact the continental crust of Laurentia. This collision took the Ordovician sandstones and shales and shoved them westward along large faults called, appropriately enough, thrust faults. As the rocks were thrust along the faults, they also internally folded due to the compressive stresses from the collision. Thus occurred the second major mountain building event to affect the Hudson Valley – the Late Ordovician Period Taconic Orogeny. Just as with the earlier Grenville Orogeny, the Taconic event was also characterized by metamorphism of sedimentary rocks and igneous intrusions. These rocks, now found on the eastern side of the Hudson River and into New England, will be discussed in the next chapter.

The Taconic Orogeny was not a simple event. The volcanic island arc was composed of several distinct blocks of continental crust, called terranes by geologists, which collided with different parts of the coast of Laurentia over time. This resulted in several pulses of mountain building affecting the eastern coast of North America from the Mid-Atlantic States northward through New England and up into eastern Canada. Where are those volcanic island terranes today? They're presently represented east of the Hudson Valley by a belt of heavily deformed igneous and metamorphic rocks over in Connecticut and Massachusetts. If you'd like a modern analogy to the Taconic Orogeny, imagine what would happen if a continent moved northward to collide with the Aleutian island arc in the northern Pacific Ocean.

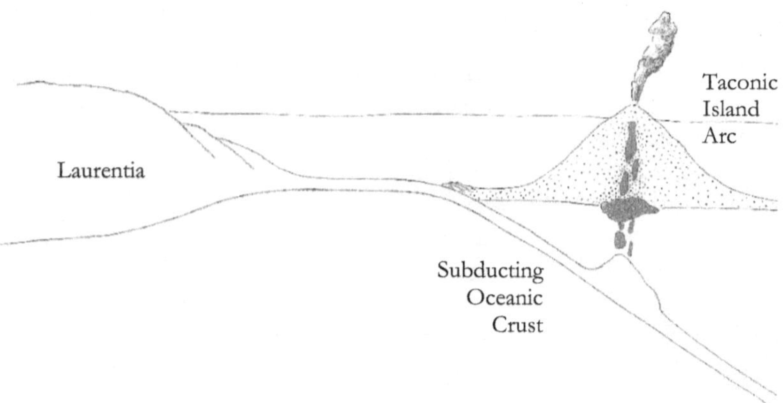

First stage of the Taconic Orogeny as the Taconic volcanic island arc approaches Laurentia as subduction consumes the intervening oceanic crust. What is now the Hudson Valley was submerged on the continental shelf of Laurentia during the early-Ordovician Period (~480 Ma).

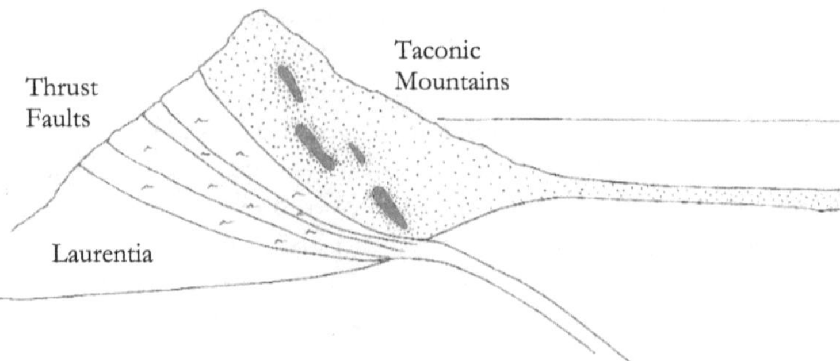

Final stage of the Taconic Orogeny in the late-Ordovician (~450 Ma) after the collision and accretion of the Taconic volcanic island arc with Laurentia forming the Rocky Mountain-sized Taconic mountains. The metamorphosed sedimentary rocks and thrust faults formed deep in these mountains are now exposed by erosion on the eastern side of the Hudson River in New York.

Today, when we walk on the heavily folded and faulted Ordovician strata of the Hudson Valley, many of the rocks we're standing on are not presently found where they originally formed. Most of these rock units formed further to our east (south of Laurentia) and were thrust westward (north toward Laurentia) by the collision. These are referred to as Taconic thrust slices. If you drive on the northern part of the Taconic State Parkway, near the Taconic Mountains in Columbia County, you're driving on a sequence of Taconic thrust slices. A more technical term for rocks that are still in their original location is autochthonous and for rocks that have been transported is allochthonous. Geologists have been arguing for the past century about whether some particular rock units are, in fact, autochthonous or allochthonous (or even parautochthonous – moved only a little bit). An additional complication for those mapping these rock units is that thrust faulting has, in some areas, emplaced older rocks directly on top of younger rocks thus "violating" the principle of superposition used to assign relative ages to different rock units. Fortunately, we don't need to worry about these details, only note that even after more than a century of study, there's still field work to be done on Ordovician rocks in the Hudson Valley.

In contrast to the sedimentary shales and sandstones in the western and northern parts of the Hudson Valley from the Ordovician Period, we'll now move to the southeastern part of the valley where these once sedimentary rocks became heavily metamorphosed during the Taconic mountain building event.

5

Manhattan's Basement

The finest garnet crystal ever found, perhaps, in the United States, was discovered, strange though it may seem, in the midst of the solidly-built portion of New York City. It was brought to light by a laborer excavating for a sewer in West 35th Street, between Broadway and Seventh Avenue, in August, 1885.

George Kunz, *New York Academy of Science Transactions*

I t's hard to imagine today, but New York City is well known among geologists for the fine mineral specimens which have been collected there over the years. The New York Mineralogical Club, for example, has a collection of hundreds of specimens currently housed at the American Museum of Natural History (including the impressive garnet crystal mentioned above). The heyday of collecting was in the late 1800s and early 1900s when numerous construction projects were excavating the bedrock of the city for subway lines, water tunnels, and skyscraper foundations. It's a little harder to collect specimens today, but the city hosts a large number of easily accessible outcrops which illustrate the wide variety of rock types to be found there.

This region of the southeastern Hudson Valley owes its geologic richness to the Taconic Orogeny. The original rocks of the area were essentially the same Cambro-Ordovician carbonates, sandstones, and shales that we discussed in the previous chapter resting on the Grenville-age crust of New York State. The difference is that from Westchester

County down through the five boroughs, these rocks were a little closer to the core of the Taconic Mountains and this made all the difference. These original sedimentary rocks were metamorphosed by this mountain building event into the marbles, quartzites, and schists present today. This area of crystalline rocks, called the Manhattan Prong, extends southward from the Hudson Highlands on the east side of the Hudson River down through most of New York City.

The oldest rock unit in the Manhattan Prong is the Fordham Gneiss. This high-grade metamorphic rock has a radiometric age of 1.1 billion years and was apparently derived from sediments and volcanic ash layers accumulating on the seafloor just prior to the Grenville mountain-building event. It's clearly related to similarly-aged gneisses found in the Hudson Highlands to the north and, for our purposes, we can consider the Fordham Gneiss to have the same geologic history as the rest of the Grenville Orogeny rocks we've already discussed.

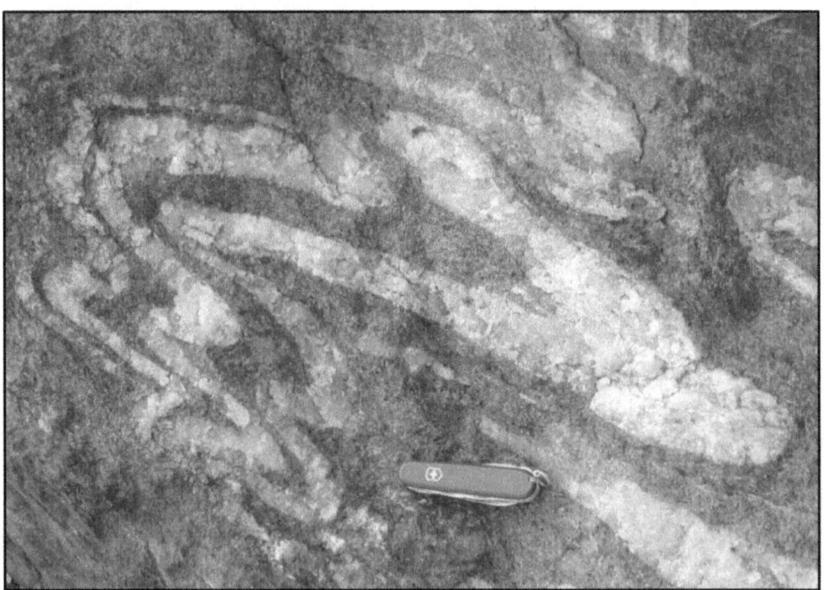

Extreme folding of quartz-rich layers of the Fordham Gneiss in Van Cortlandt Park

The Fordham Gneiss is a hard rock which is resistant to weathering and thus forms ridges throughout the area where it's exposed in a belt through Westchester County, southward into the Bronx, and on the northern tip of Manhattan. Excavations have also shown that it's present in the subsurface of Queens and Brooklyn as well. Outcrops of Fordham Gneiss are seen in innumerable road cuts in southern Westchester

County. Two examples are along Interstate 684, especially just south of Exit 4, and near the eastern approach to the Tappan Zee Bridge on the Thruway. Unfortunately, they can only be studied while sitting in a traffic jam since you're not allowed to park and walk along these roads. If you would like to see the Fordham Gneiss up close, you can climb all over it on the large set of outcrops behind the Parade Ground in Van Cortlandt Park in the Bronx. The high grade of metamorphism, along with the extreme folding commonly seen throughout gneissic layers within the rock, attest to this rock's formation deep within the Grenville Mountain belt at extremely high temperatures and pressures.

There are other gneissic rocks exposed in Westchester County besides the Fordham Gneiss. Two of the better known are the Pound Ridge Gneiss and the Yonkers Gneiss. These rocks can, in places, be hard to distinguish from the Fordham Gneiss but radiometric dating of small grains of the mineral zircon ($ZrSiO_4$) within them has shown that they are half a billion years younger and formed a little over 560 million years ago. These gneisses apparently represent granitic intrusions into the Fordham Gneiss basement rock around the time Rodinia was rifting apart and starting to form the Iapetus Ocean.

This rifting may have been initiated by a mantle plume, a jet of hot material rising from deep within the mantle forming a "hot spot" on the Earth's surface. Such a mantle plume exists under the present day Big Island of Hawaii in the middle of the Pacific Plate. As the Pacific Plate moved northwestward over the past few tens of millions of years, the entire chain of Hawaiian Islands (and numerous seamounts) formed as they passed over this hot spot.

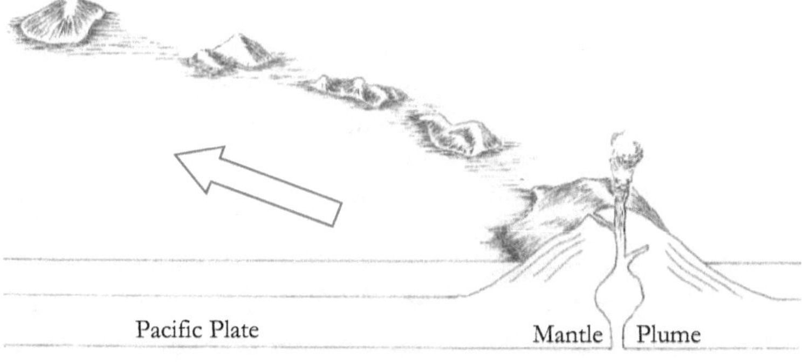

Formation of the Hawaiian Islands by movement of the Pacific Plate over a hot spot

A number of mantle plumes have been identified around the Earth today forming other volcanic provinces and islands including locations beneath continental crust. One is believed to lie under Yellowstone

National Park, for example, which besides being characterized today by hot springs and geysers was also the site of a catastrophic volcanic eruption a mere 600,000 years ago. A mantle plume forming under the thick, thermally-insulating crust of Rodinia over 600 million years ago would explain why it rifted apart where it did to form the coast of Laurentia. It would also explain why granitic magmas were generated around this time in our area since heat from the mantle plume would have melted the granitic continental crust of Rodinia. The bodies of magma would have then risen through the crust and cooled to form new masses of granitic rocks which were later metamorphosed into granitic gneisses.

There are a number of other rock units dating from this time period, when rifting was tearing apart Rodinia. In the Manhattan Prong area, geologists have referred to these as the Ned Mountain Formation and they represent a complex package of metamorphosed igneous and sedimentary rock units from the Late Proterozoic Eon (about 570 million years ago). These rocks, however, can be difficult for the non-expert to distinguish from the underlying Grenville-age rocks.

Once a passive margin developed on the edge of Laurentia, marine sediments began to accumulate. Remember the Cambrian Period Poughquag Quartzite and Wappinger Group dolostones exposed in Dutchess County? These same types of sediments accumulated here as well. The only difference is that south of the Hudson Highlands, these sedimentary rocks were more heavily metamorphosed by the Taconic Orogeny. Actually, even though I spoke about the Poughquag and Wappinger Group rocks as sedimentary in Chapter 3, they were slightly metamorphosed in Dutchess County as well, just not to the same degree as rocks south of the Highlands (we'll come back to this shortly).

Metamorphism (derived from Greek for "change in form") is the process by which preexisting rocks are altered by high temperatures and pressures into new rocks with different physical properties from their original protolith (Greek for "first stone"). Textbooks define the metamorphic field as ranging in temperature from 200° C to 800° C (roughly 400-1,500° F). Below 200° C, rocks are essentially unaltered. Above 800° C, rocks begin to partially melt since different minerals have different melting temperatures. Rocks, however, may not be completely melted into magma until temperatures of 1,000° C are reached. How much rocks are altered by metamorphism depends on the types of minerals comprising the rock, the highest temperature reached during metamorphism, and the overall heating and cooling history of the rock. Pressure also plays a role in metamorphism as does water circulating through the rock and exchanging chemical ions.

When a sedimentary quartz sandstone is metamorphosed, for example, the resultant metamorphic rock is called a quartzite. Quartz sandstones are composed of rounded quartz sand grains held together by a mineral cement, typically itself composed of quartz (which geologists call silica cement). Sandstones have lots of pore space between the grains and rubbing a quartz sandstone will often dislodge them. Sandstones essentially erode back into the sand grains from which they originally formed. During metamorphism, however, these grains of sand within a sandstone will fuse together forming a much harder, more resistant rock. The pore space disappears and when a quartzite breaks, it will often break right through quartz grains rather than around them as they do in sandstone.

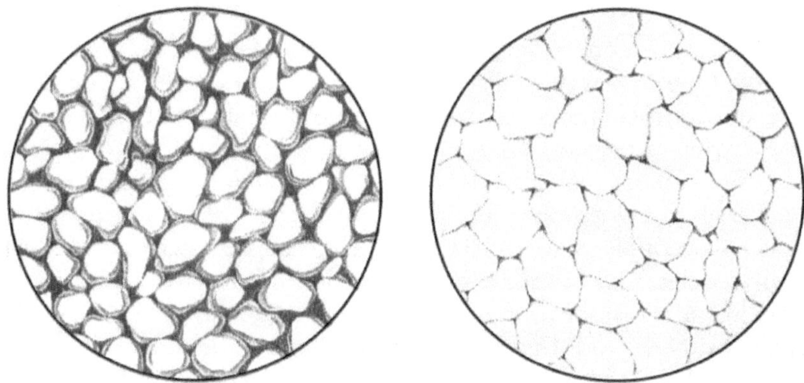

Microscopic view of cemented quartz grains in a sedimentary sandstone (left) and intergrown quartz crystals in a metamorphic quartzite (right).

Sedimentary carbonate rocks such as limestone will also alter during metamorphism. The carbonate mud, calcite grains, and fossil fragments within these rocks will reform into intergrown calcite crystals when subjected to high enough heat and pressure. The resultant rock is called a marble and slightly different types of marbles will form depending upon the composition of the protolith limestone. A sandy (arenaceous) limestone may have some quartz sand mixed in, a clay-rich (argillaceous) limestone may have some clay mud mixed in, and a dolostone has a significant proportion of magnesium mixed in with the calcium carbonate. These different compositions will each result in different types of minerals forming within the resultant marble during metamorphism as we'll soon see.

In the Manhattan Prong, two units called the Lowerre (pronounced "low-wear") Quartzite and the Inwood Marble correspond to the Poughquag quartz sandstone and Wappinger Group dolostones exposed in Dutchess County. The Lowerre Quartzite lies on top of the Fordham Gneiss (Grenville crust) and represents the early appearance of the Sauk marine transgression. The Inwood Marble represents the carbonate shelf that followed and persisted on the Laurentian passive margin from the Cambrian into the early Ordovician Period.

The Inwood Marble is an important and well known rock unit in the New York City area. It's best exposed in the southern Bronx and northern Manhattan but has a larger extent in the subsurface (it's often encountered during excavations). It also correlates to a rock unit called the Stockbridge Marble across the border in Connecticut. The Inwood is termed a dolomitic marble due to its magnesium content which is not surprising given that it corresponds to the Wappinger Group dolostones. The marble is quite attractive in outcrop, being white in color with a sugary texture where it's been exposed to surface weathering.

Calcite, the dominant component of marble, is a relatively soft mineral. One easy way to distinguish quartz from calcite, if you don't have any hydrochloric acid handy, is to do a scratch test. Quartz will easily scratch a piece of glass while calcite will not. Or, if you carry a pocketknife, most geologists do, a steel blade will not scratch quartz but it will easily gouge a groove into a piece of calcite. Calcite, unlike quartz, is also susceptible to chemical weathering. Slightly acidic water will dissolve calcite (it's simply a weaker version of the hydrochloric acid effervescence mentioned in a previous chapter). Rainwater is naturally acidic, and New York City rainwater even more so due to air pollution, so exposed marble will dissolve over time. That's why the Inwood Marble is generally found underlying valleys rather than ridges. The Harlem area, for example, is a low area underlain by marble while Washington Heights, on the western side of Manhattan overlooking the Hudson River, is underlain by the more resistant Manhattan Formation (discussed soon).

One place to conveniently examine an outcrop of Inwood Marble is in Isham Park at the corner of Seaman Avenue and Isham Street in the northeastern part of Manhattan Island. Gray-colored siliceous (quartz-rich) layers along this outcrop stand out in slight relief due to their greater hardness and likely reflect thin sandstone layers present in the original dolostones. If you look carefully, you'll also see that these rocks have been deformed by folding, a topic we'll discuss later in this chapter.

The Inwood Marble is an interesting unit and has long been used as an attractive building stone. In northern Manhattan, near Isham Park, marble was quarried as early as the 18th century for downtown buildings like Federal Hall on Wall Street, the first capital of the United States where George Washington was inaugurated. From the early 1800s to the early 1900s, the Tuckahoe Marble Quarry in Westchester County provided stone for many of the neo-classical buildings and monuments throughout the City. The Inwood Marble also occurs further up the Hudson River near Ossining where it was used to construct the infamous Sing Sing Prison (the phrase about being "sent up the river" referred to this institution which once housed an electric chair for carrying out the death penalty). The Sing Sing Prison quarry, operational back to when prisoners really did break rocks all day, was conveniently located near the Hudson where marble could be shipped down to Manhattan or up to Albany where it was also used for many building projects (New York University in Manhattan and the State Capitol in Albany, are two of many examples).

The Sing Sing Prison quarry is also famous among mineral collectors because of the numerous fine-quality mineral specimens which have been discovered there. Slight impurities in the original dolostone protolith were altered by chemical reactions during metamorphism to form these minerals in the resultant Inwood Marble. The types of minerals which form depend not only upon the chemical elements available within the rock but also the temperatures reached during metamorphism since different minerals are stable at different temperatures, an important fact we'll discuss in more detail soon. Many Sing Sing Prison quarry minerals are now housed in the New York State Museum in Albany and the American Museum of Natural History in Manhattan.

In Dutchess County, north of the Hudson Highlands, the early Ordovician was marked by a marine regression which left the Knox Unconformity on top of the Wappinger Group carbonates. We then see the subsequent appearance of the thick sequence of Ordovician shales and sandstones of the Normanskill Group. In the Manhattan Prong, these sedimentary rocks were also metamorphosed by the Taconic mountain-building event and the resultant group of rocks is generally referred to as the Manhattan Formation. Unfortunately, as with the Ordovician Period rocks to the north, the details are complicated and a little messy. In the ensuing discussion, I will be following the general interpretation of these rocks published by Hofstra University professor Charles Merguerian and others who have done extensive field work in the New York City area.

Let's travel back to the Cambrian and Ordovician time periods. During the Cambrian Period, the passive margin continental shelf along the south coast of Laurentia is gradually giving way to a deeper anoxic basin as the Taconic island arc terrane approaches. Carbonates are forming in the shallow shelf waters and characterized by the Wappinger Group dolostones (Dutchess County) and the Inwood Marble

(Westchester County). Further offshore, however, sandstones and shales are forming in the deeper waters nearer the trench. As the trench approaches, water depths increase in our area and the environment here changes during the Ordovician Period to one where shales and sandstones are now accumulating in the Hudson Valley Region. Then the Taconic arc collides. Three different packages of rock form in the Manhattan Prong area and all become heavily metamorphosed, primarily into schists.

Schists are medium- to high-grade metamorphic rocks with a very distinctive texture formed from overlapping scales of mica minerals. To understand the formation of schists, it will be helpful to explain the concept of progressive metamorphism.

Imagine a shale. Shales are simple sedimentary rocks composed primarily of clay minerals – the end result of the chemical weathering of minerals from preexisting rocks. As mentioned in earlier discussions, shales are essentially lithified mud and form in a variety of environments where muds accumulate. When examined under a microscope, the platy clay minerals in shale are roughly parallel to each other which impart a heterogeneous texture to the rock. In other words, shales have parallel planes of weakness and will split relatively easily along their bedding planes into small chips.

As you move down into the crust, temperatures increase at roughly 25° C per kilometer of depth. This is called the average geothermal gradient. If a shale is buried deeply enough such that it begins to metamorphose (> 200° C), clay minerals within the shale will become unstable and alter into mica. Mica denotes a whole group of minerals but two important types are muscovite, or white mica, and biotite, or black mica. While differing in color, both micas have similar physical properties – they can be peeled apart into thin flexible sheets. Because of its resistance to heat, white mica (which is actually clear when peeled thin enough) has been used to make windows for furnaces (the name muscovite actually comes from "Muscovy glass" given its former use in the city of Moscow). Micas are also good electrical insulators and are commonly used in electronic components called capacitors.

The specific type of mica which forms from the metamorphism of a shale is dependent upon the chemical makeup of the original clay minerals in the protolith. These in turn depend on the types of minerals which originally weathered into clays. The first type of metamorphic rock formed from the thermal alteration of shale is called slate. In a slate, the micas formed from the original clay minerals within the shale are microscopic and cannot be seen with the naked eye.

In addition to heat from the geothermal gradient, a deeply buried shale will also experience pressure as it's being metamorphosed. This pressure will align the flat, microscopic flakes of mica parallel to each other resulting in what's called slaty cleavage. The property of slaty cleavage is what allows slates to be split into large, flat sheets of rock. The heat and pressure of metamorphism also results in slate being typically much harder than shales. You can often snap apart shale chips with your bare hands, slate requires a hammer and chisel.

Slaty cleavage makes slate very useful for a number of purposes. It's commonly used for chalkboards (although markers and whiteboards have taken over our classrooms today), pool tables, roofing tiles, and patio pavers. Slates are found in the Hudson Valley, primarily east of the River and north of the Hudson Highlands where Normanskill Group shales were slightly metamorphosed by Taconic mountain-building. Numerous slate quarries are located along a belt of outcrops in the northern Taconics straddling the New York/Vermont border.

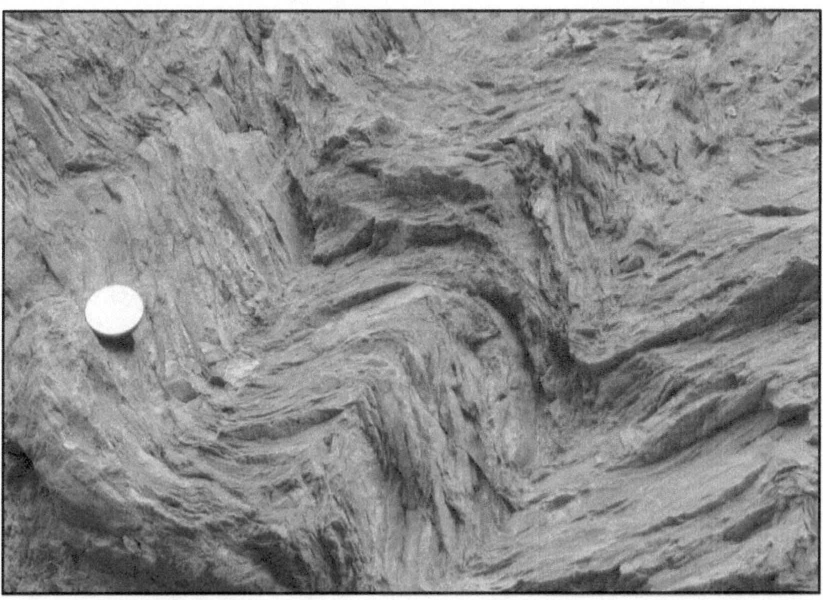

Extreme folding in phyllite along Noxon Road in Poughkeepsie (quarter for scale)

Slate, however, is only a low-grade metamorphic rock formed at modest temperatures and pressures. As temperatures increase, the microscopic mica minerals in slate grow larger to a size where they're barely seen by the naked eye. At this size, the mica crystals impart a shine to the rock which is commonly called a satin sheen (the rock reflects light much like a piece of satiny fabric). These new metamorphic rocks are called phyllites and are also seen in the Hudson Valley. Outcrops of spectacularly-folded phyllites, again derived from metamorphosed Normanskill Group shales, are found just east of Poughkeepsie on Noxon Road near the Taconic State Parkway.

We're not done, however, since phyllites can still metamorphose further as temperatures and pressures continue to climb. With increasing heat, the subvisible micas in phyllites will grow into larger visible flakes. At these medium- to high-grade metamorphic conditions, we now form a schist. While the dominant minerals in schist are sparkling flakes of micas, they also contain some quartz as well as other minerals that form during metamorphism. One common metamorphic mineral is garnet, although there are many others that can occur (and the term garnet actually refers to a whole group of specific types of garnets with different chemical compositions). Geologists typically distinguish between different schists by listing their most abundant mineral components as a descriptor before the rock name (e.g. muscovite-garnet schist). A number of different schists are found in the eastern Hudson Valley region north of the Highlands but we'll confine our following discussion to those found south of the Highlands in the Manhattan Prong.

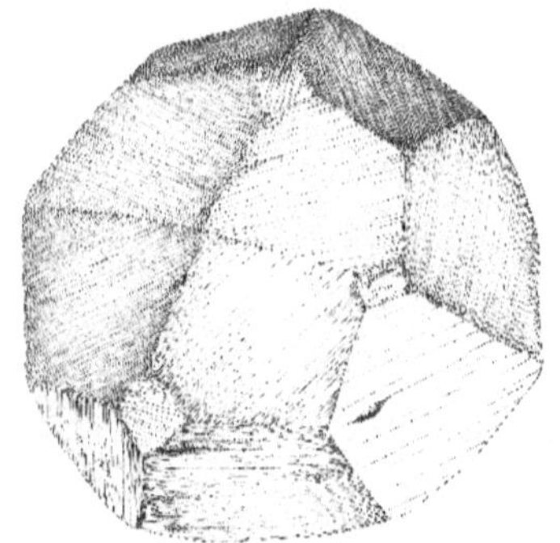

Cantaloupe-sized (18 cm) garnet crystal mentioned in the epigraph to this chapter

Let's return to those three packages of rock that formed in the New York City area from the Taconic Orogeny. The first package of rock is called the Walloomsac Schist which lies on top of the Inwood Marble and is interpreted to be Ordovician in age. The Walloomsac is thought to have formed in place (remember the term autochthonous?) and would essentially correlate with the Martinsburg and Normanskill Groups of sedimentary shales and graywacke turbidites found to the north. The Walloomsac Formation is not well exposed at the surface and often difficult to distinguish in the field.

The second package of rock is the famous Manhattan Formation (often called the Manhattan Schist). The Manhattan Formation is composed of dark-colored gneisses, schists, and amphibolites that are well exposed throughout Manhattan and up into Westchester County (with outliers as far north as Pawling in eastern Dutchess County). Amphibolites are metamorphic rocks composed primarily of the mineral hornblende (a black mineral belonging to the amphibole group) and they may have formed either from the metamorphism of volcanic sediments derived from the Taconic volcanic arc terrane or through the metamorphic alteration of basaltic oceanic crust. The protolith for the Manhattan Schist apparently formed in deeper water during the Cambro-Ordovician Periods at the same time the carbonate rocks were forming in our area. In other words, they represent rocks from the deeper water continental slope and rise while the carbonates were accumulating on the continental shelf. They now lie on top of the Walloomsac Schists, which are younger, because they were thrust over them during the Taconic collision like the analogous Normanskill Group thrust slices east of the Hudson River in the northern part of the Valley.

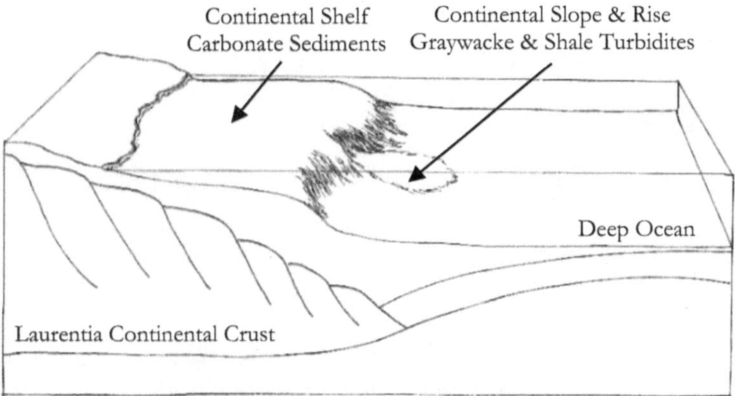

Continental margin of Laurentia. Sedimentary rocks which formed in these different environments were later thrust onto Laurentia and metamorphosed into marbles and schists during the collision of the Taconic island arc.

The Manhattan Formation is easy to find. A stroll through Central Park is probably the best way to examine this rock up-close and personal. The mica-rich schist layers form an attractive rock that sparkles in the sunlight although weathering often turns the exposed outcrops rusty-red due to the oxidation of iron-bearing minerals in the rock. Outcrops sometimes show glacial polishing and striations as well, but that's a story for a later chapter.

The Manhattan Formation is a relatively hard rock and forms ridges in Manhattan. The Washington Heights area of northwestern Manhattan is located on this rock unit. Inwood Park, across the street from the marble outcrops in Isham Park mentioned earlier, is a great place to explore these rocks as well. The Manhattan Formation, as with all the metamorphic rocks of the Taconic Orogeny, is heavily folded. The modern-day skyline of Manhattan Island is a reflection of this subsurface structure. Skyscrapers need a strong foundation and clusters of these buildings occur downtown in the Financial District (site of the former World Trade Center) and in Midtown (think of the Empire State building) where the Manhattan Formation is near the surface. Skyscrapers are lacking, however, in Greenwich Village which is between these two areas because the Manhattan Formation dips down to about 80 meters (260 feet) below ground level in this area and the overlying glacial deposits formerly made constructing foundations for towering buildings here economically impractical (although modern building techniques have mitigated this a bit).

The third and final package of rock is called the Hartland Formation. These are also schists and gneisses but have a slightly different composition than the Manhattan Formation. Geochemical analysis of this unit suggests that it's roughly the same age as the Manhattan Formation but may have originated in deep waters on the opposite side of the Taconic arc trench. Its protolith was also deep-water shales and turbidites with a lot of volcanically-derived sediments and it's also allochthonous – it was thrust over younger rocks by the Taconic collision. The Hartland Formation is primarily found in the eastern part of the Bronx, Queens, eastern Westchester County, and maybe southern Manhattan (still open for debate since it may be entirely different from the Hartland Formation elsewhere!).

There is a very significant geologic boundary between the Hartland Formation and the Manhattan Formation called Cameron's line that was first recognized by geologist Eugene Cameron in the 1950s. Cameron's line is a fault that essentially marks the ancient boundary between two tectonic plates – the Laurentian Plate and the plate which contained the Taconic island arc volcanoes on the other side of the subduction trench. While the Walloomsac and Manhattan Formations formed as part of Laurentia, the Hartland Formation represents what geologists call an exotic terrane – a block of crust which formed elsewhere and was accreted to what's now the North American continent through ancient plate tectonic collisions. This is the story of all the modern continents – they're patchwork quilts of different tectonic terranes which grew over time through tectonic collisions.

Cameron's line is actually difficult to see in the field and most easily recognized by regional mapping of rocks in the Manhattan Prong. The consensus is that it runs southward through eastern Connecticut, crosses into Westchester County near Rye Lake, roughly follows the Bronx River down to the East River where it makes a complicated S curve, and then travels through Queens, Brooklyn, and into Staten Island. Everything east of Cameron's line was not part of North America prior to the Late Ordovician Period of geologic time.

While the Cameron's line fault is an ancient structure, some 450 million years old, there are a number of other faults crisscrossing the city which occasionally wake from their slumbers and shake the area. The greater New York City area has experienced more than a dozen light to moderate earthquakes in historic times. One of the largest, with an estimated 5.2 magnitude, occurred in August of 1884 off Far Rockaway in Queens and was felt throughout the Eastern Seaboard. Another estimated 5.2 magnitude earthquake occurred earlier in December of 1737 just across the Hudson from Manhattan in New Jersey.

Faults are often found during subway and water tunnel excavations, they're generally not visible on the surface. One can, however, sometimes identify their location by the fact that faults are often areas of increased erosion because the rock there is already fractured and broken. This increased rate of erosion will leave a trace on the landscape over time and a number of northwest-to-southeast trending linear features are seen cutting across Manhattan Island.

One of the more famous fault traces cutting Manhattan is the 125th Street (Manhattanville) fault. This fault runs diagonally from the area around 125th Street and Broadway near the Hudson River on the west side of the island and from there through the Harlem Meer (a lake on the northeastern end of Central Park) and finally crossing the East River around 96th Street and FDR Drive. It presumably extends westward into New Jersey and eastward into Long Island as well. While, thus far, it has only been the site of small earthquakes, like the magnitude 2.4 trembler in January of 2001, there's no reason why it couldn't generate larger earthquakes in the future.

Earthquakes in the eastern United States are called intraplate earthquakes because they occur in the middle of a tectonic plate and far from any plate boundaries. It's easy to understand why earthquakes occur in California. The relentless movement of the Pacific Plate as it grinds past the North American Plate along the San Andreas Fault system generates seismic waves of energy as rocks periodically snap and slip.

Earthquakes occur in the East (and in the Midwestern U.S.), however, because preexisting faults which may have formed hundreds of millions or even billions of years ago, are reactivated by stresses transmitted thousands of kilometers into the interior of the continental plate from interactions along the plate boundaries.

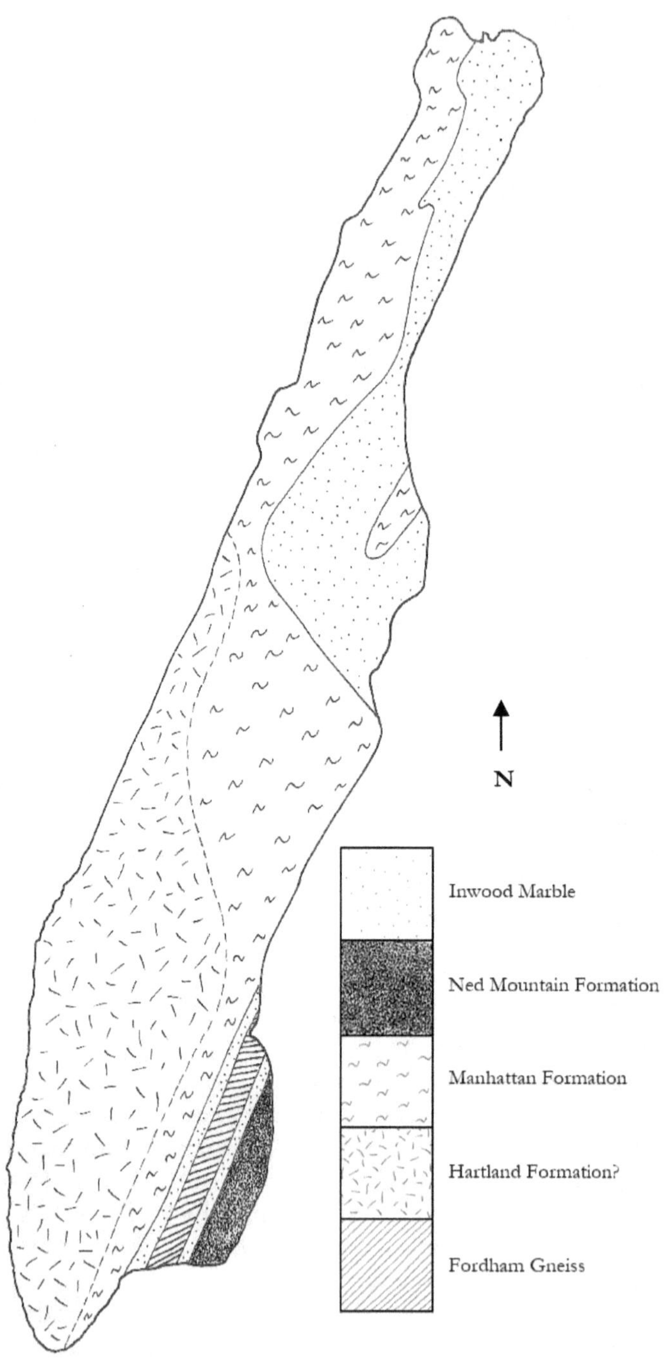

Simplified geologic map of Manhattan Island.

Outside of the greater New York City area, earthquakes have also been recorded in the Hudson Valley along the Ramapo Fault. The Ramapo Fault, actually an intertwined system of faults, is the boundary between the Hudson Highlands to the north and west and the Rockland County lowlands (west of the Hudson) and Manhattan Prong (east of the Hudson) to the south and east. As with the 125th Street fault, these relatively small (thus far) earthquakes represent modern-day reactivation of an ancient geologic structure. Hopefully, neither of these large fault systems will experience larger earthquakes since the results could be catastrophic for a city not generally built to withstand such shaking. While the Indian Point nuclear power plant was actually cited very near to where the Ramapo Fault crosses the Hudson River, it was built to withstand a magnitude 6 earthquake. Unfortunately, recent studies of historical seismicity in the area suggest that magnitude 5 earthquakes may occur as frequently as once every 100 years and magnitude 6 earthquakes are theoretically possible, albeit rare, possibly occurring once every 600-700 years (and we don't know when the last one might have shaken the area).

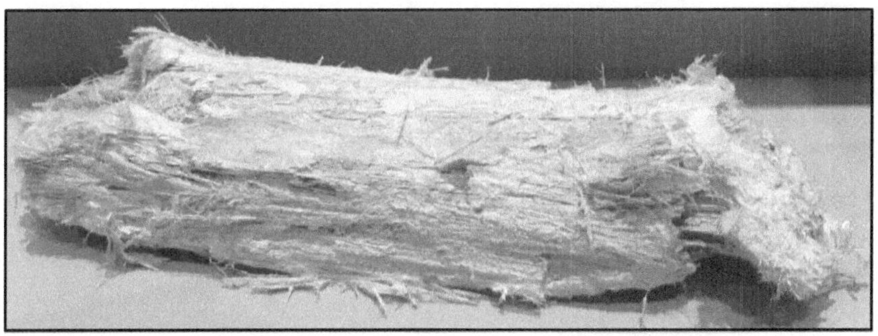

Anthophyllite (asbestos) specimen in the New York State Museum at Albany collected in Staten Island

If anyone needed more convincing that a collision between Laurentia and a chain of volcanic islands occurred during the Late Ordovician Period, one only has to visit Staten Island where a large outcrop of serpentinite forms the center of the island. Serpentinite, the official state rock of California, is a metamorphic rock derived from an igneous rock called peridotite. Peridotite is the dominant rock type in the mantle of the Earth (the mantle, below the crust, is 80% of the Earth's volume) and composed of crystals of an olive-green silicate mineral called, appropriately enough, olivine $((Mg,Fe)_2SiO4)$. When olivine is metamorphosed in the presence of hot water, a process called hydrothermal alteration, it forms a group of minerals collectively called serpentine. One well-known serpentine mineral forming this way is chrysotile asbestos; a mineral infamous for causing mesothelioma lung cancer (asbestos has been commercially mined in Staten Island since the late 1800s).

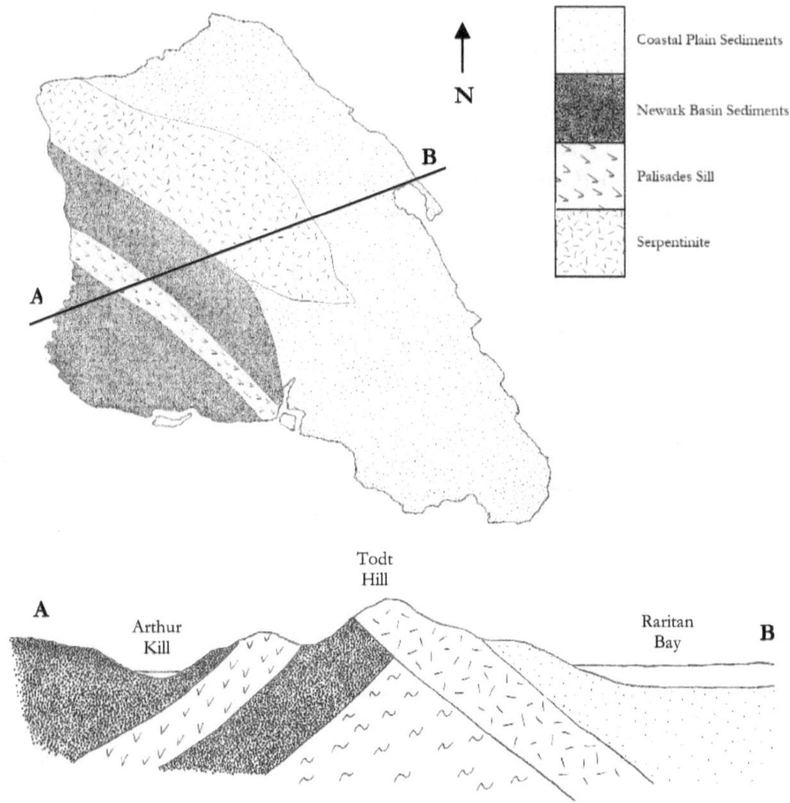

Geologic map and cross section of Staten Island

So where did this peridotite come from, how was it metamorphosed into serpentinite, and why is it sitting in the middle of Staten Island? It turns out that pods of serpentinite are common in areas where subduction zones collide with masses of continental crust (which is why they're found in places like California where this process has occurred as well). Just below the relatively thin basaltic crust of the seafloor is mantle peridotite. When the Taconic island arc began to collide with Laurentia, small slices of the seafloor crust and uppermost mantle were chopped up and thrust over together with the allochthonous thrust slices of the Hartland Formation (like the Hartland Formation, this seafloor crust was moving along the Cameron's Line fault zone). During Taconic mountain building, hot groundwater circulating through these rocks metamorphosed them into serpentinite.

When you stand on the serpentinite, you're essentially standing on Ordovician Period seafloor crust. If you'd like to experience this thrill, exposures on Todt Hill, just south of the Staten Island Expressway (Interstate 278), are a famous field trip area for northeastern geologists. It may be hard to believe, but this hill is the highest point along the Atlantic seaboard south of Maine at a modest 125 meters (410 feet) above sea level. Smaller pods of serpentine are also found elsewhere within the New York metropolitan area. One such pod was identified under CUNY's John Jay College of Criminal Justice in Midtown Manhattan, for example.

A final Ordovician rock unit that deserves a mention is the Cortlandt Complex in northern Westchester County. This rock unit outcrops along the east side of the Hudson River just south of the City of Peekskill. The Cortlandt Complex is relatively small, covering only about 65 square kilometers (25 square miles) of land surface, but is geologically significant. It represents several intrusions of magma into the Cambro-Ordovician rocks of the Manhattan Prong that occurred around 430 million years ago near the end of the Taconic mountain-building event. It appears to be related to other intrusions of roughly the same age extending from Connecticut to New Jersey.

Remember those late-stage regional faults and magmatic intrusions that occurred in the Hudson Highlands after the Grenville mountain-building event due to "escape tectonics" and which are associated with the magnetite iron ore deposits? A similar sequence of events likely resulted in these Ordovician-age intrusions deep within the Taconic Mountains. Interesting minerals are present within some of these unusual igneous intrusions including emery which is commercially-mined near Peekskill. Emery is a rock comprised mostly of corundum ($Al2O3$), an extremely hard mineral valued for its abrasive properties. Two other varieties of corundum, not found in the Hudson Valley, are rubies, colored red by small amounts of chromium, and sapphires, colored blue by small amounts of iron and titanium. The emery deposits near Peekskill formed when blocks of the surrounding Manhattan Schist fell into these intruding magmas and were metamorphically altered by the extremely high heat into emery.

The Taconic Orogeny took about 20 million years from start to finish between 460 and 440 million years ago. Like the earlier Grenville Orogeny, it likely occurred in phases and the resultant mountain range didn't pop up overnight. The metamorphic rocks we've discussed in this chapter formed within this mountain belt at depths around 40 kilometers (nearly 25 miles) or more. At these depths, the rocks would have experienced high-grade metamorphic temperatures close to their melting point and pressures around 12 kilobars (almost 12,000 times the atmospheric pressure at the Earth's surface). Today, these high-grade metamorphic rocks are exposed at the surface because, as with the earlier Grenville Orogeny, the original mountains beneath which they formed have long since weathered away.

The Taconic Mountains stretched from the Carolinas to the Canadian Maritimes and their jagged, snow-capped peaks would have been an impressive sight in the Late Ordovician Period. While it's difficult to work out the height of mountains that haven't existed for hundreds of millions of years, the degree of metamorphism and subsequent amount of sediments eroded from them suggest that parts of the range certainly exceeded 4,000 meters (13,000 feet) in elevation. This places them in the same class as the present-day Rocky Mountains.

By the close of the Ordovician Period, the Hudson Valley region was dry land on the western side of the Taconic Mountains. There are some tantalizing fossils of possible spores from other localities suggesting that plants may have started to colonize land by the end of the Ordovician, but they are few and far between. The land surface was still essentially barren of life at this time with high rates of erosion. This period of mountain building and erosion, lasting for tens of millions of years, resulted in a widespread unconformity in our area between the heavily deformed Cambro-Ordovician rocks and younger rocks deposited on top of them. This boundary is called, not surprisingly, the Taconic Unconformity. Our story will next pick up above this boundary in the Silurian Period of geologic time.

6
Rivers of Pebbles

Again, do you not behold stones too vanquished by time, high towers falling in ruins, and rocks crumbling away... And stones torn up from high mountains rushing headlong, unable to brook or bear the stern strength of a limited time? For indeed they would not be suddenly torn up and fall headlong, if from time everlasting they had held out against all the siege of age without breaking.

Titus Lucretius Carus, *De Rerum Natura*

One of my favorite places in the Hudson Valley is the Shawangunk Ridge in Ulster County. The Shawangunks, locally pronounced "shon-gums," are a great place for hiking with interconnected trails running from the Mohonk Preserve near Rosendale southward through Lake Minnewaska State Park and on to Sam's Point Preserve near Ellenville – a straight line distance of about 18 miles, much more on the trails. On a nice day, sitting on the edge of an exposed ridge of white rock amid gnarled pitch pines or walking through shafts of sunlight in a mossy, cathedral-like forest of hemlocks is truly a sublime experience. The hard rock of the Shawangunks, or "Gunks" as some so unflatteringly call them, are world famous among rock climbers who are often seen "thick as pigeons" (to borrow a phrase from the late Russell Waines, a former SUNY New Paltz geology professor) on the cliff faces near Route 44/55 on the east side of the ridge.

The Shawangunk Ridge lies between the Rondout Valley on the west, separating the Shawangunks from the Catskills; and the Wallkill Valley on the east, separating the Shawangunks from the Hudson. The Shawangunks first appear near the town of Rosendale and run southwestward as a relatively narrow ridge just east of Route 209. Following the Delaware River, between Pennsylvania and New Jersey, the ridge runs along the east side of the river where it's known as the Kittatinny Mountains. High Point State Park, the highest spot in New Jersey, is located along this segment. After passing through the Delaware Water Gap at Interstate 80, the ridge extends southwestward into Pennsylvania where it's called the Blue Mountains.

Rock shelters and archaeological artifacts indicate that native Lenape Indians were hunting in the Shawangunks over 10,000 years ago and many of the modern routes over the ridge, such as Route 44/55, follow old Indian trails. One of the earliest European mentions of the Shawangunks I'm aware of is a reference from mid-17th century Dutch explorers who spoke of a "crystal mountain" west of the Hudson River – presumably referring to how the white quartz-rich rock sparkled in the sunlight. As settlers spread through the Hudson River area, the Shawangunk Ridge formed a natural barrier between the Wallkill and Rondout valleys.

While tracts of the Shawangunk Ridge are now protected, especially in Ulster County, today's ever-encroaching development is just the continuation of a long history of exploitation of this area by European settlers. Though much of the ridge is unsuitable for farming due to the thin, acidic, rocky soil covering the bedrock, abandoned homesteads from the 18th and 19th centuries are common. It's not unusual to come across networks of stone walls delineating old fields and even remnant building foundations deep in the forests. Many of these historic sites can be found in the Mohonk Preserve in the northern part of the Shawangunks.

One industry that developed in this area, due to the extreme hardness of the rock which forms the ridge, was the manufacture of millstones. When the Hudson Valley was first being settled, almost every community had a grist mill to grind locally-grown grain into flour. Mills originally imported their stones from Europe (which came over as ship ballast), but locally-quarried stone, called Shawangunk Grit historically, soon became preferable. There are a number of abandoned quarries in the Shawangunks and millstones can typically be seen throughout the wider area as large, decorative lawn ornaments. It always amazes me, when looking at a millstone, that someone carved these large round stones from granite-hard rock using only simple hand tools. I once calculated the weight of a millstone blank found near an old quarry in the Mohonk

Preserve at 1,260 kg (over 2,700 pounds)! This round blank, 1.4 meters (4.5 feet) in diameter and 30 centimeters (1 foot) thick, was carved from an even larger slab of rock which had already been transported out of the small quarry – all with muscle power.

Shawangunk Conglomerate millstone in town of High Falls

The hardness of the local rock along with its abundant quartz has led some locals to believe that it's a type of granite. For many years, a resort hotel near the town of Kerhonkson on the flanks of the Shawangunks was called the Granite, and its present incarnation as the Hudson Valley Resort is still located on Granite Road. I've also heard local well drillers refer to this rock unit as granite when they encounter it in the subsurface. The rock that makes up the Shawangunk Ridge is not granite, it's not even an igneous rock (remember that granite forms when molten rock cools and solidifies deep in the Earth's crust). The Shawangunks are composed of conglomerate, a sedimentary rock made up of rounded pebbles and quartz sand cemented together by a chemically-precipitated quartz cement (called silica cement by geologists).

If you're ever fortunate enough to find yourself hiking in a mountainous area, take some time and examine the sediments you find in any swiftly-flowing stream. Odds are that you will find plentiful sand within the streambed along with rounded rocks of various sizes ranging from small pebbles up through larger cobbles (rocks the size of your fist). These stones are modified as they're transported in streams because random collisions with each other will preferentially chip off their projecting corners resulting in a rounded rock over time. Take these streambed sediments, bury them such that they can lithify into sedimentary rock, and you will end up with conglomerate.

Pebbles in the Shawangunk Conglomerate

When we left off our story at the end of the previous chapter, the majestic Taconic Mountains were looming over the Late Ordovician landscape of what's now the Hudson Valley. The Ordovician rocks in the western part of the Hudson Valley were primarily the folded and thrust-faulted shales of the Martinsburg Group. These rocks were folded and faulted at depth during the Taconic mountain-building event. How do we know they were folded at depth? Pick up a piece of shale and try to fold it. It won't work, the shale will simply snap. This is called brittle deformation. For plastic deformation (keep in mind our discussion of shear zones in Chapter 2), the rock needs to be heated up and placed under high pressure. This can only occur when the rock is buried a few kilometers down in the crust.

For over 20 million years, from the Late Ordovician into the early part of the Silurian Period, our area was a terrestrial environment dominated by erosion. While a few primitive plant fossils have been found from Silurian rocks worldwide, they had yet to make any significant progress far inland and were likely confined to moist shoreline and riverbank environments. One example of a Silurian plant fossil found in New York State is *Cooksonia*. Several different species of this primitive vascular plant have been identified around the world but they are all similar in appearance. *Cooksonia* has a number of simple branching stalks, only a few centimeters tall, with each branch topped by a rounded spore case. They had no leaves and photosynthesized through their green stems.

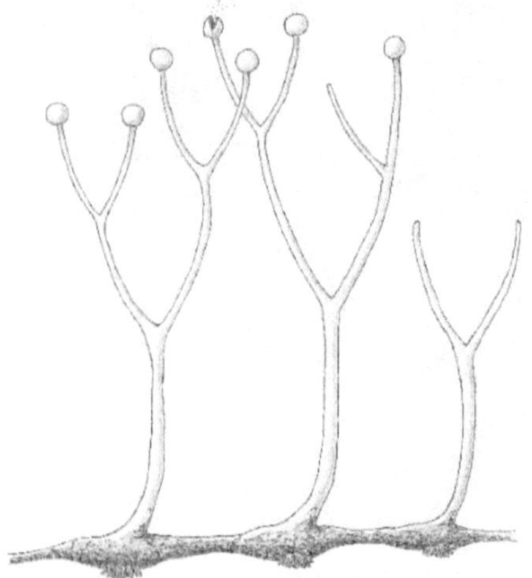

Silurian plant reconstruction

Mountain streams were rushing down from the Taconics, turning into braided, sediment-rich rivers on coastal floodplains, and then flowing westward into a shallow sea covering what's now the central and western parts of New York State. This landscape would have been mostly barren, perhaps with a few scattered green patches of primitive plants, and covered by sediments. Erosion had brought the folded and faulted Ordovician Period shales and sandstones to the surface and these stream-deposited sediments accumulated on top of them. This erosional surface formed a prominent contact we now call the Taconic unconformity. We've already discussed nonconformities which are formed when sedimentary rock is deposited on igneous or metamorphic rock as with the Poughquag Formation on top of Grenville rocks. We've also discussed disconformities which are formed when sedimentary rock is deposited on top of sedimentary rock after a hiatus in deposition as with the Knox unconformity on top of the Wappinger Group. The Taconic unconformity, however, is called an angular unconformity and it formed when horizontal sediments were deposited on top of deformed sedimentary rocks. The name obviously refers to the angular relationship between the two layers of rock and it represents a gap in time since folded rocks form at depth in the Earth's crust, not at the surface. For the folded rocks to be exposed such that new sediments can be deposited on top of them, time is needed for erosion to occur.

Taconic angular unconformity near Catskill. Diagonal contact separates almost vertical Ordovician shales and sandstones from tilted Silurian dolostones. A long period of erosion is represented by this contact.

The Taconic unconformity is easy to find and unmistakable in eastern New York State. While this contact originally formed as a horizontal layer, subsequent mountain-building events, which we'll discuss later, have folded even this contact. Between the Late Ordovician rocks deformed by Taconic Orogeny mountain-building and the Late Silurian rocks originally deposited horizontally on top of them, is a layer of unconsolidated sediments that has been interpreted as a paleosol – an ancient soil. In other words, the Silurian rocks were deposited on top of a terrestrial erosional surface.

Throughout its length, the Shawangunk Formation conglomerates and sandstones are sitting on top of the Taconic angular unconformity. The Shawangunk Formation is primarily composed of rounded, white quartz pebbles and quartz sand grains that are very strongly bound together by silica cement. These sediments form interbedded conglomerates (pebble-rich layers) and sandstones (sand-rich layers).

The Shawangunk sandstone layers are very well-cemented, resembling the metamorphic rock quartzite, and are often referred to as orthoquartzites. Some of the quartz pebbles, especially near the upper part of the formation, are actually pink rose quartz. A few minor accessory minerals are found within the unit as well as some thin red and green shale layers. The Shawangunk Formation is actually quite complex in detail and is subdivided into different members along the length of the ridge. The Shawangunk conglomerate also grades into a very similar rock unit called the Tuscarora Sandstone in the Blue Mountains of Pennsylvania to the south.

The source area for the Shawangunk Formation sediments were higher elevation areas to the east and a large contribution was likely from pushed up Hudson Highlands rocks. As the Taconic Orogeny pushed allochthonous blocks of Ordovician rocks westward, it also thrust up the more ancient basement rocks of the Grenville Province along faults on the northern edge of the present-day Hudson Highlands. As these rocks became exposed due to the erosion of the overlying sedimentary rocks, they in turn began to erode to provide sediments for the Shawangunk Formation.

There are several good pieces of evidence for this inference. First, the source area had to be rich in quartz since that's what comprises virtually all of the Shawangunk Formation sediments, and Grenville-age gneisses have abundant quartz. One problem with this, however, is that some geologists wonder if they have enough quartz to provide the amount we see in the Shawangunks. Second, the Shawangunks pinch out and disappear in the mid part of the Hudson Valley as do the Hudson Highlands. If the source area for the sediments were the Taconic Mountains proper, we would expect the Shawangunks to extend into the northern Hudson Valley. And last, but most importantly, there are zircon crystals that have been found in the Shawangunk Formation that radiometrically date from between 950-1200 million years old placing them squarely in the range of ages we would expect for zircons eroded from the Grenville-age gneisses of the Hudson Highlands.

This is a good example of how historical sciences like geology work. We don't know for sure exactly where these ancient sediments came from. We may never know. But, given evidence (e.g. the radiometrically-dated zircon crystal), we can make a reasonable hypothesis. Future evidence may further support or even disprove this hypothesis. We always have to be willing to change our ideas based on the physical evidence. This is why it annoys many scientists when we're accused of not having an open mind — especially when we refuse to give credence to nonsense like young-Earth creationism, dowsing, or crystal healing. We have open minds, but only to ideas that are supported by empirical evidence.

Besides the character of its sediments, the Shawangunk Formation also contains other evidence attesting to its fluvial, or stream-deposited character. Imagine a barren plain, sloping steeply out of the high mountains and covered with a network of braided stream channels. These types of stream systems exist today in some environments (Alaska, for example) and are often described as being "a mile wide and a foot deep." Braided streams are characteristic of areas where there is an abundant supply of water and a lot of sediments such as you might find near large mountain ranges.

The streams would contain lag deposits of rounded pebbles lining the channels – sediments too large to be picked up under normal flow conditions. Larger channels might have sandy beds with ripples on the sediment surface formed from the flowing water. In between the braided channels would be constantly shifting sand and gravel bars. Sheets of sand would lie on the floodplain. Spring snowmelt in the Taconic Mountain would send down torrents of water while late summer droughts would practically dry up the area. And, to top it all off, this was a dynamic system where the positions of the channels and bars would be constantly changing over time (sometimes from day to day).

Pebble-rich stream channel layer in the Shawangunk Conglomerate changing vertically into a cross-bedded quartz sand-rich layer. Quarter for scale.

The features seen in soft sediments in this type of environment can be preserved when the sediment lithifies into solid rock. They're then called sedimentary structures and geologists value them for what they can tell us about the ancient environment in which the sediments accumulated. Fluvial environments have a number of very distinctive types of sedimentary structures and we see these in the Shawangunk Formation. We see trough-shaped channels of pebbly conglomerate embedded within surrounding sandstone – ancient stream channels. We see a feature called trough cross-bedding within the sandstone layers which forms from the migration of ripples in the sand beds of larger streams. And we see the lateral migration of these cross-bedded sands and channel conglomerates over time as we examine these sedimentary structures throughout the extent of the Shawangunks.

One thing that's lacking in the Shawangunk Formation, however, is fossils. Fluvial environments typically don't preserve fossils very well, especially during the Silurian Period when there simply wasn't much available for fossilization in the way of terrestrial life. Only a few rare fossils have been found in the Shawangunks. One type is the enigmatic trace fossil *Arthrophycus alleghaniense*, which is thought to represent the tubular cast of an invertebrate feeding burrow. While we don't know the exact organism that excavated these burrows, these fossils are found elsewhere in the world in transitional areas between fluvial and a marine environments. In other words, these organisms lived in estuaries where rivers flowed into the sea.

Another rare group of fossils found in the Shawangunk Formation are eurypterids – an extinct class of arthropods sometimes called "sea scorpions." While most eurypterid species were less than 30 centimeters (1 foot) long, a few grew as large as 2 meters (over 6 feet) in length. Eurypterids existed throughout the Paleozoic Era, like the trilobites, but they were most important as top predators in the Silurian seas of New York State. In recognition of this, the species *Eurypterus remipes* was named the official state fossil in 1984. Most of the Silurian eurypterids of New York are found in the western part of the State, especially within a rock unit called the Bertie Dolostone exposed in a belt between southern Herkimer County and Buffalo. A few eurypterid fossils, however, have also been discovered in the Shawangunk Formation such as *Hughmilleria shawangunk*.

Eurypterids were similar to modern-day horseshoe crabs (*Limulus polyphemus*) in their mode of life. While primarily marine, eurypterid fossils are typically found in sedimentary rocks from transitional environments – the coastal area between land and sea. Eurypterids as a group apparently tolerated a wide variety of salinity conditions and are found in areas ranging from highly-saline seawater to brackish lagoons to freshwater rivers. Also, like horseshoe crabs, eurypterids were likely able to walk around on dry land for short periods of time and could have been occasionally seen on the beaches, river banks, and tidal flats of the Paleozoic world. Why were they venturing out onto land? They most likely made this transition because food sources were starting to appear in the terrestrial environment due to the evolution of land plants. Plants attract herbivores and herbivores attract carnivores like the eurypterid.

Silurian eurypterid reconstruction

Let's now place the formation of the Shawangunk Formation into a larger regional picture of what's going on in the rest of New York State since it will help us to understand the next stage in the evolution of the Hudson Valley.

During the Silurian Period (one of the shorter Paleozoic Era subdivisions), Laurentia was still rotated almost 90° from its present-day position and was mostly in the Southern Hemisphere. Only the western part of what's now the United States was north of the equator. We were in a subtropical, yet arid, climate zone — it was hot and dry.

This brings us back to the Tippecanoe marine transgression, mentioned in Chapter 4, which started in the Late Ordovician Period and covered most of New York with the Trenton Sea and forming the Trenton Group limestones west of the Hudson Valley. Our area has no marine rocks from this time because we were an uplifted area closer to the Taconic Mountains. It's the same story all over again, recalling the quotation which began this chapter, as we "…behold stones too vanquished by time." The Taconic Mountains, which formed during the Late Ordovician Period, are now eroding away throughout the 28 million years of the Silurian Period.

As the Taconics eroded, sediments were transported westward by streams and rivers and began to fill the Trenton Sea with sands and mud. Some estimates indicate that over 600,000 cubic kilometers of rock (imagine a cube over 50 miles on each side) were eroded and transported as sediments into this sea. This type of environment, a shallow inland sea between a mountain belt and the continent, is called a foreland basin by geologists and foreland basins typically fill with sediments over time forming thick sequences of sedimentary rock in the geologic record.

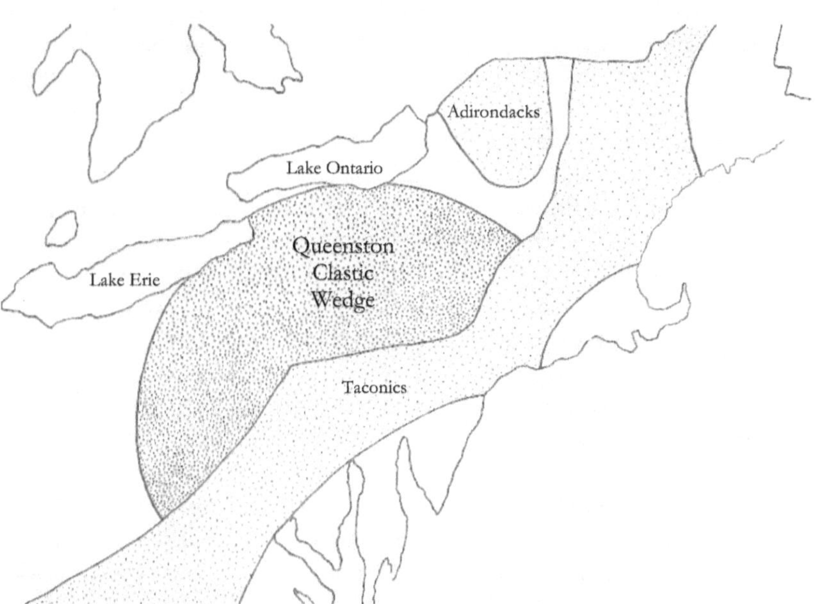

The Queenston clastic wedge is a thick pile of sediments (pebbles, sand, silt, and clays) eroded from the ancient Taconic Mountains during the Silurian Period.

The sediments that filled in the Trenton Sea were derived from the Taconic Mountains further to the east during the latter part of the Ordovician Period (prior to the deposition of the Shawangunk Formation). These sediments were composed of sands and clays which formed thick deposits of sandstone and shale within the foreland basin. Sediments, derived from the physical and chemical weathering of preexisting rocks are often referred to as clastics (from a Greek root meaning "broken"). Clastic sediments formed a wedge-shaped Late

Ordovician deposit (called a clastic wedge) which was thickest in central New York and then thinned to the west extending as far out as Ohio. Parts of this wedge are almost 4,000 meters (13,000 feet) thick. This deposit is known as the Queenston clastic wedge and is named after rocks which outcrop just over the Niagara River in Queenston, Ontario. One impressive place to see Queenston clastic wedge sediments (the eroded remnants of the once-mighty Taconic Mountains) is within a deep gorge at Whetstone Gulf State Park in the Tug Hill Plateau region of New York (an area between the Adirondacks and Lake Ontario).

The sediments of the Queenston clastic wedge completely filled in the Trenton Sea. By the end of the Ordovician Period, these rocks were exposed to terrestrial weathering which removed an unknown amount of sediment from the top of this sequence. With erosion, and the continued rising of sea levels due to the Tippecanoe transgression, the area once again become flooded with seawater. A sequence of shallow marine sandstones, shales, and carbonate rocks, some richly fossiliferous, was once again deposited throughout western and central parts of New York State during the Silurian Period.

By the Late Silurian Period, while the Shawangunk Formation was being deposited in a fluvial environment in the Hudson Valley Region, western New York was characterized by a shallow sea teeming with coral reefs. If you would like to see an example of the rock unit forming in this environment, simply visit Niagara Falls. The massive gray rock which caps the falls is the Late Silurian Lockport Dolostone. If this rock wasn't so hard and resistant to weathering, the Niagara River would gradually lose elevation in a series of rapids and not abruptly drop over the edge of a picturesque waterfall.

Coral reef structures preserved as fossils within the Lockport Dolostone are significant because they tell us so much about the ancient environment in this part of New York during the Late Silurian Period. Coral reefs today are generally found only within 30° north or south of the equator since coral polyps (the actual animal forming coral) require warm ocean water. When we look at coral reef distributions in the geologic past, we see a similar pattern. So while coral isn't found off the northeastern coast today, New York State was within 30° of the equator during the Silurian Period and its warm, shallow seas teemed with corals.

Reefs have changed over geologic time. The earliest reefs in the geologic record were composed of stromatolites (remember them?). With the evolution of marine invertebrate herbivores after the Cambrian Explosion, stromatolites became rare and reefs were often composed of calcareous sponge-like organisms. One important group of reef builders which existed only during the Cambrian Period, for example, were archaeocyathids. The name means "ancient cups" and they were small double-walled, cone-shaped structures composed of calcite (like corals) and punctuated by tiny holes. It's the holes that indicate their affinity to sponges, which belong to phylum Porifera because of these pores. While individual archaeocyaths were only a couple of centimeters long, collectively they built impressive reef structures many kilometers in length. Even though they both arose and became extinct in the Cambrian Period, one shouldn't think of them as unsuccessful since they were around for 10-15 million years (far longer than hominids like us have been on Earth).

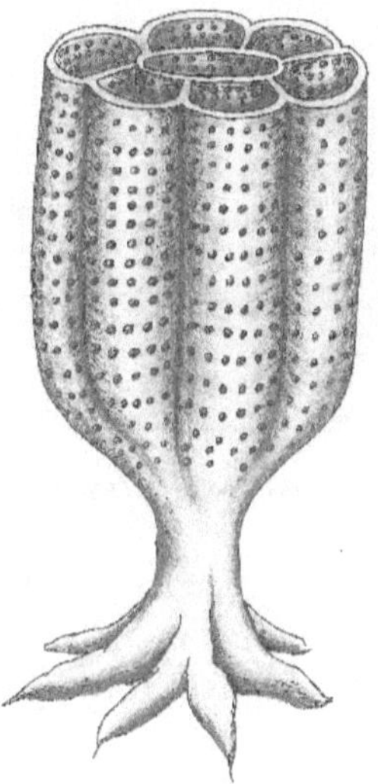

Reconstruction of a Cambrian archaeocyathid

It was only after the Late Ordovician extinction event, discussed earlier, that corals started to become widespread in the geologic record. While all corals belong to the same taxonomic phylum (Cnidaria) and class (Anthozoa), there are three important groups which have existed throughout geologic time. The modern corals we are familiar with today are called scleractinian ("hard-rayed") or stony corals and have dominated the world's oceans since the early Mesozoic Era. Scleractinians replaced two earlier groups, now extinct, that filled the Paleozoic seas – rugose and tabulate corals. Rugose ("wrinkled") corals, also called horn corals, were primarily solitary organisms and formed curved, ridged, horn-shaped corallites. These corallites are very common as fossils in some of the rock units in the Hudson Valley as we'll discuss later. Tabulate corals, on the other hand, were colonial and had interior platforms, called tabulae, upon which the coral polyp was located.

Tabulate coral fossils are also common in many of the sedimentary rock units of the Hudson Valley – clear indicators that our area was once covered by a shallow, warm sea.

Top and side views of a colonial tabulate coral (left) and a solitary rugose coral (right)

While coral polyps, the animal living within each corallite, are only a few millimeters to several centimeters in length, they developed the remarkable ability to extract ions from seawater to secrete calcium carbonate ($CaCO_3$) and collectively construct some of the largest biological structures on the planet. The Great Coral Reef system off Australia, for example, stretches for 2,600 kilometers (1,600 miles) and can be seen from space. Since corals are composed of calcium carbonate, and can be quite sizable, they are commonly preserved as fossils in limestones around the world. We'll discuss some specific examples of Hudson Valley corals shortly.

Above the Lockport Dolostones in western New York, and dating from the end of the Silurian Period, is a sequence of sedimentary rocks belonging to the Salina Group. The name Salina is derived from the Latin word for salt and this group of rocks is well-known for its deposits of the minerals halite and gypsum. Halite is what geologists call mineral deposits of salt ($NaCl$) and gypsum is hydrated calcium sulfate ($CaSO_4 \cdot 2H_2O$). Both are economically important – the salt is mined primarily for use on icy New York roads and gypsum is used in the manufacture of wallboard. Since these minerals form by the evaporation of seawater, they're often termed evaporite minerals.

Remember that our area had a hot and arid climate during the Silurian Period. The Tippecanoe transgression was now waning and sea levels were dropping around the world. Shallow basins of seawater, restricted from circulating much by fringing coral reefs, had high evaporation rates depositing thick beds of these evaporite minerals from New York State westward into Michigan. Some of these salt deposits have cumulative thicknesses of hundreds of meters and would have required the evaporation of massive volumes of seawater. Today these salt deposits are buried in the subsurface beneath younger sedimentary rocks and recovered by either underground mining or by the injection of water through deep wells, dissolving the salt beds, and then pumping out and evaporating the resultant saline brines at the surface.

Finally, atop the Salina Group is the Bertie Group consisting mostly of dolostone formations. The Bertie is world-famous for containing richly abundant fossils of eurypterids. One locality, called Passage Gulf and located just south of the city of Herkimer, has yielded large numbers of spectacular specimens over the past few decades. Another famous eurypterid locality is located near Buffalo and fossils from these localities can be found in museums throughout the world. The end of deposition of the Bertie Group in western New York represents a drying up of the area (a marine regression) and a hiatus in deposition until the Middle Devonian Period.

Returning to the Hudson Valley, there are a number of excellent places where you can view the Shawangunk Formation in Ulster County. Hiking trails in the Mohonk Preserve or adjacent Lake Minnewaska State Park will provide breathtaking views westward to the Catskills and eastward across the Hudson Valley to the distant Berkshires as you walk along exposed ridges of white rock. The many cliffs in the park afford an opportunity to examine the rocks in detail and reveal the pebbly layers of conglomerate along with the cross-bedded sandstones we discussed earlier. Notice the horizontal layering which is called bedding in sedimentary rocks. Bedding planes form due to subtle differences in the deposition of sediments – factors such as variations in sediment influx, fluctuations in water depths, and changes in current flow directions or velocities. Sedimentary beds are often planes of weakness within the rock unit and may pop open when the rock is exposed at the surface thus providing a conduit for groundwater flow. Shortly after a rainstorm, exposed cliffs will often have water flowing out of these open bedding planes, and layers of icicles are usually present during the winter months.

In addition to the bedding plane fractures, cliffs also exhibit sets of vertical fractures called joints. These are very important in how Shawangunk cliffs erode over time. Water easily seeps into open joints and multiple cycles of freezing and thawing each winter effectively work to open up the joint over time. Water expands by about nine percent when it freezes and can exert a tremendous force on the side walls of a fracture. This physical weathering process eventually releases large blocks of rock which fall and accumulate at the base of cliffs as talus. Vertical joints often form in two parallel sets intersecting each other at approximately right angles. It's not uncommon to see a sawtooth pattern on cliff faces resulting from this phenomenon.

Talus at the base of cliffs at Lake Minnewaska State Park

There are several ways to form joints in sedimentary rock. One type forms due to brittle fracturing of the rock as it's folded in the subsurface. If the rock layer is folded a bit too much, or a little too fast, tensile stresses will fracture the rock, typically in areas where the fold has its maximum curvature. Joints associated with folding are local features and are easy to identify given their geometric relationship with the fold.

Unloading is a mechanism by which regional sets of joints can form in a rock layer. When rocks are in the subsurface, they're under a lot of confining pressure from the weight of the rock overburden above. When these layers are later eroded to the surface, the exposed rock will physically expand and pop open parallel sets of joints, often at right angles.

Regional joints may also form by the release of tectonic stresses. During a collisional orogeny, rocks are subjected to large horizontally-directed forces in the crust which impart a compressive stress. At the end of the mountain-building event, these stresses relax and the rock expands popping open fractures perpendicular to the former stress direction.

All three of these mechanisms have operated in the Shawangunks resulting in the heavily jointed cliff faces seen throughout the ridge. While the Shawangunk Formation has virtually no open pore space on a microscopic level, this three-dimensional network of joints allows for the storage and migration of groundwater throughout this unit.

Bedding plane and vertical joints in Shawangunk Conglomerate cliffs — black vulture (Coragyps atratus) for scale

At Sam's Point Preserve, near Ellenville, jointing has resulted in neat features known as ice caves. Large joints developing on the edge of a cliff formed massive blocks of rock which tilted outward and slid a bit on the weaker layer of underlying Martinsburg Group shales. These large crevices, in some places roofed over with other slabs of rock, form talus caves. In the winter, snow accumulates in these crevices and often doesn't melt until late summer (if at all). While presently owned by the Nature Conservancy, this area was once commercially developed and called Ice Caves Mountain. These ice "caves" are still accessible to anyone willing to hike a few miles and I would highly recommend visiting them. There is another set of crevices on the other side of the ridge from the formerly commercial ice caves, sometimes called the "greater ice caves," but these ecologically sensitive features are more protected and require a local guide to find and safely explore.

Joints, along with other geologic structures such as folds and faults, are found throughout the Shawangunk Ridge but did not form until a mountain-building event that occurred well over 100 million years after these sediments were deposited. This is another key point to keep in mind when discussing the geology of the Hudson Valley region – the formation of rock units and the deformation of rock units does not usually occur at the same time. Rocks may exist for a long time before they're affected by later geologic events. We'll discuss the deformation of the Shawangunk Ridge when we talk about the Alleghenian Orogeny and the formation of Pangaea in Chapter 10.

One of the more interesting geologic features of the Shawangunks, again related to its more recent geologic history that we'll discuss later, are deposits of minerals rich in lead and zinc. Two of the most common minerals are galena, a lead sulfide (PbS), and sphalerite, a zinc sulfide (ZnS). Crystals of galena and sphalerite are usually found growing on a matrix of beautiful quartz crystals along with accessory minerals such as pyrite (FeS2), chalcopyrite (CuFeS2), and others. These mineral deposits formed within joints and fault zones in the conglomerate by chemical precipitation from hot, mineral-rich waters circulating through the fractures. Mining, primarily in the 19th century, resulted in the collection of hundreds of mineral specimens, many of which have made it into collections at the State Museum in Albany and the American Museum of Natural History. The importance of Shawangunk lead-zinc mining is reflected by the fact that Route 209, just west of the Ridge, was once known as the Old Mine Road.

Geochemical analysis of Shawangunk mineral deposits seem to indicate that they formed during the Mesozoic Era while the supercontinent of Pangaea was breaking apart, a topic we'll discuss in Chapter 11. This rifting event mobilized hot groundwater flow from deep basement rocks which migrated up through the fracture systems formed previously by the Alleghenian Orogeny. Unfortunately, most of the old Shawangunk mines are either filled in, flooded, or otherwise inaccessible today. I've only visited a few of the mines, including a historically interesting tunnel just outside of the city of Ellenville. The tunnel, extending some 90 meters (300 feet) straight into the mountain, is inexplicably known as the Old Spanish Mine after a spurious legend that it was dug by exploring Conquistadors (there's no credible evidence that wandering Spaniards ever entered the Hudson Valley prior to the modern era). A permanent spring results in a year-round stream running out of this mine and prior to World War II, the Sun Ray Bottling Works sold this as "the world's purest spring water." Since this mine is on private property and is a sensitive bat hibernaculum in the winter months, I won't reveal its exact location.

While hiking in these areas, the rocks aren't the only interesting thing to observe (is it blasphemy for a geologist to say this?). There are a number of unique ecosystems in the Shawangunks, which is one of the reasons that people seek to preserve the ridge from increased development. A number of battles have taken place over the years by local citizens who have opposed the building of large-scale resorts in the heart of the Shawangunks – a place the Nature Conservancy once called "One of America's last great places."

The thin, acidic soils along the crest of the Shawangunk Ridge host a large number of pitch pines (*Pinus rigida*). These are easily distinguished from other pines by their clusters of three, stiff needles and small, prickly cones. Chestnut oak (*Quercus prinus*) and white paper birch (*Betula papyrifera*) are scattered amongst the pitch pines and large tracts of delicious low-bush blueberries (*Vaccinium augustifolium*) cover many areas of the ridge top. These blueberries are called huckleberries by old-timers in the area and were historically picked in large quantities from the ridge around Sam's Point Preserve. Mountain laurel (*Kalmia latifolia*) and sheep laurel (*Kalmia augustifolia*) are also ubiquitous and bloom with spectacular clusters of pink flowers in early June, a prime time for hiking the trails.

The slopes of the Shawangunks, mostly underlain by Martinsburg Group shales, are characterized by mossy glens and ravines in perpetual shade from towering hemlocks (*Tsuga canadensis*), white pines (*Pinus strobus*), and the occasional massive hardwood. There are very few old-growth hemlocks remaining in the Shawangunks due to the former use of hemlock bark (which contains tannin) in the tanning industry. Hemlocks in the Catskills, an area we'll discuss in a later chapter, were also extensively logged for the same reason

While now protected from logging, Shawangunk hemlocks are today endangered by an introduced Asian pest, the wooly adelgid (*Adelges tsugae*). Wooly adelgids, named for their cottony egg masses found on the underside of Hemlock needles, are tiny, aphid-like insects that weaken the tree when large numbers of them feed on sap at the base of the tree's needles. If too many needles are lost, the tree may die and it's not uncommon to see the sad, skeletal remains of such trees throughout the area.

As we reach the end of the Silurian Period, the environment which produced the Shawangunk Formation conglomerates and sandstones begins to change as shallow seas once again begin to encroach upon the Hudson Valley area.

The earliest indications of this transgressing sea are red shales which show signs of being deposited in a muddy tidal-flat environment. These are followed by marine sandstones and carbonate units. In the northern part of the Shawangunks, in Ulster County, these units are called the High Falls Shale, Binnewater Sandstone, and Rondout Formation. Generally equivalent beds further south where the Shawangunk Ridge extends into New Jersey and Pennsylvania are called the Bloomsburg Red Beds and Poxono Island Formation.

These latest sedimentary rock units record the earliest appearance of the Helderberg Sea in our area. This is one of the more important rock-forming events in Hudson Valley geology and the subject of our next chapter.

7
The Helderberg Sea

Thus, evidence which cannot be rebutted, and which need not be strengthened, though if time permitted I might indefinitely increase its quantity, compels you to believe that the earth, from the time of the chalk to the present day, has been the theatre of a series of changes as vast in their amount, as they were slow in their progress. The area on which we stand has been first sea and then land, for at least four alternations; and has remained in each of these conditions for a period of great length.

Thomas Henry Huxley, *On a Piece of Chalk*

Thomas Huxley, once nicknamed "Darwin's Bulldog" for his spirited defense of Charles Darwin's theory of evolution, was one of Britain's leading intellectuals of the mid-19th century. One of his most admirable qualities, in my opinion, was his commitment to public education and the popularization of science. The quotation above is from a lecture Huxley gave to a working-class audience in Norwich, England where he showed how you can reconstruct the geological history of Britain by common-sense reasoning from an ordinary piece of native chalk.

It's my hope that this book, although a bit more long-winded than Huxley's erudite lecture, will similarly show how we can reconstruct the geologic history of the Hudson Valley from simple observations of local

121

outcrops. Granted, much of modern geology now relies on sophisticated technology to examine rocks under high magnifications, geochemically analyze their composite elements, radiometrically date those we can, and develop complex computational models; but these are just refinements and additional verifications of the story we can derive from simple observations and logical reasoning.

People are sometimes surprised when they learn that the Hudson Valley was once covered with seawater. They're only surprised because they've never taken the time to carefully look at the bedrock beneath their feet. Once you've seen limestones containing fossil reefs and other marine invertebrates, it's easier to believe than if someone just tells you it's true. That's why it's so important for students to go out and look at actual exposures of rock. Field work is an indispensable component of a geological education and if you want to really learn about the geology of our area, I strongly recommend you take some time and visit the areas mentioned in this book. Don't just drive past rock outcrops at 60 mph, get out and look at them up close. What you find in the most mundane of locations may surprise you.

There's a very well-developed sequence of rocks exposed in the Hudson Valley which date from the Late Silurian up through the Middle Devonian Periods of geologic time – a span of about 40 million years. These rocks clearly record several marine transgressions in our area – one of which is called the Helderberg Sea. This sea was quite extensive stretching from New York State southward into Virginia. Limestones formed in this sea are well exposed along the Helderberg Escarpment, a prominent east-facing cliff in Albany County (Helderberg is derived from the Dutch for "bright mountains"). The escarpment has been a classic field site for geologists since the early 1800s and is now protected in John Boyd Thacher State Park.

As the Taconic Mountains weathered away throughout the Silurian and into the Devonian Period, the area covered by seawater migrated northward and eastward over time toward the eroding highlands. By the Early Devonian Period, much of eastern North America was submerged. As the influx of clastic sediments slowed and then ceased, limestones became the dominant type of sedimentary rock forming in this sea. Not all limestones, however, were created equally. Geologists can recognize many different types of limestones which formed under different paleoenvironmental conditions (e.g. shallower water, deeper water, clay influx from nearby land masses, etc.).

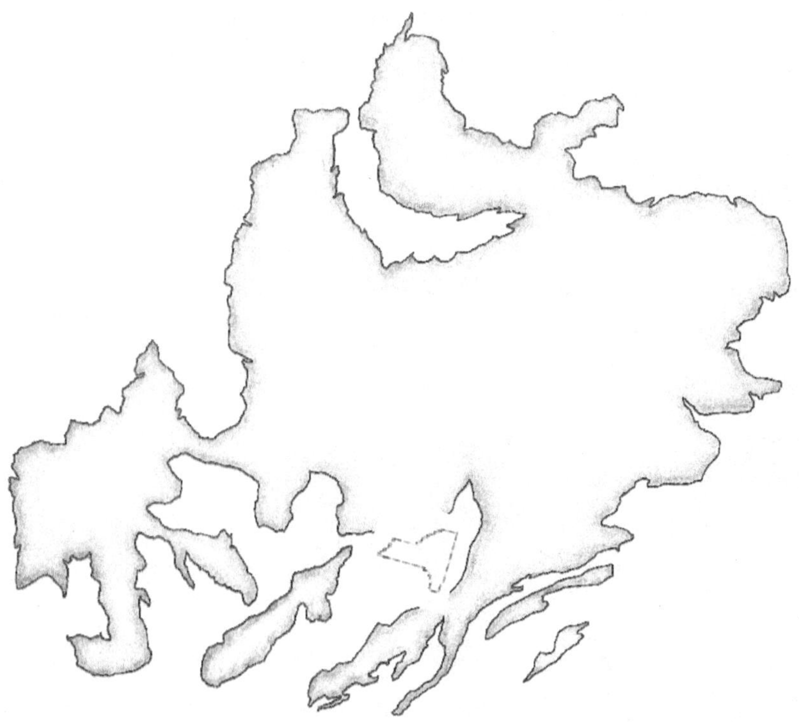

Reconstruction of Laurentia during the Early Devonian Period (~410 Ma). Approximately 2/3 of the continent was south of the equator. The approximate location of modern-day New York State is shown in the Helderberg Sea.

As we've already discussed, the Shawangunk Formation represents a Late Silurian fluvial (stream) environment. Rivers flowing off higher areas to the east were carrying rounded pebbles and sands toward the inland seas of central and western New York. As the mountains eroded, the shoreline migrated eastward and we soon see the deposition of muddy tidal-flat sediments followed by overlying shallow marine sands. In the northern Shawangunks this is represented by the High Falls Shale and Binnewater Sandstone and in the southern Shawangunks by the Bloomsburg Red Beds and Poxono Island Formation. In the Hudson Valley, the next rock formation above these units is the Late Silurian Rondout Formation.

The Rondout Formation is composed of four distinct members that you can learn to distinguish between in the field. From bottom (oldest) to top (youngest), they are the Wilbur Limestone, the Rosendale Dolostone, the Glasco Limestone, and the Whiteport Dolostone. Note the alternation between limestone (calcium carbonate) and dolostone (calcium-magnesium carbonate) which likely reflect fluctuations in sea level. Sediments forming in a tidal flat environment, probably hypersaline, formed dolostone. Evidences for a more hypersaline tidal flat environment for the dolostones are a paucity of marine invertebrate fossils as well as preserved mud cracks in many places. On the other hand, the limestone units were deposited in slightly deeper offshore waters and host a number of fossil species attesting to their marine origin. The different members can be distinguished from each other in the field by their different reactions to hydrochloric acid (limestone effervesces more than dolostone) as well as the dolostone units containing small amounts of iron which cause them to weather to a buff-brown color while the limestones weather gray.

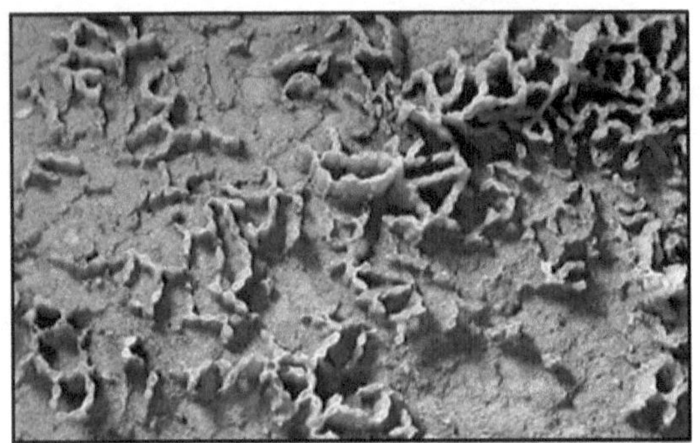

Photo and reconstruction of Cystihalysites catenularia

Two important types of fossils in the Rondout Formation, especially visible in the Glasco Limestone Member, are stromatoporoids and chain corals. Stromatoporoids, not to be confused with the stromatolites we've discussed earlier, were a type of calcareous sponge composed of flat sheets (laminae) separated by short vertical pillars. They're now extinct but were important reef builders in the early Paleozoic Era. In outcrops, stromatoporoids often look like wavy, cabbage-leaf structures in the limestone. During the process of fossilization, the original calcite material in the organism was replaced by silica, a much harder mineral, so that the fossils will weather out of the softer rock in relief. The other type of fossil characteristic of the Glasco Limestone is commonly called chain coral (*Cystihalysites catenulatus*) due to its unmistakable appearance. Coral polyps lived within each of the chain links. The Rondout Formation is placed in the Silurian Period because *Cystihalysites* was extinct by Devonian times. Additional coral species, brachiopods, trilobites, and other marine invertebrates are found fossilized in these rocks as well.

The Rosendale and Whiteport Dolostone Members are also very

important in the geological history of the Hudson Valley because these units can be used to produce a high-quality natural cement. A large number of abandoned mines and cement kilns from the City of Kingston southwestward into the Rondout Valley through the Town of Rosendale stand silent witness to this once booming industry. Since much of the natural cement was produced in the Rosendale area, it came to be known far and wide as Rosendale cement. The story of Rosendale cement started with a fortuitous set of circumstances in the early 1800s.

In 1825, construction began on the privately-financed Delaware and Hudson (D&H) Canal, a 108 mile route from Honesdale, Pennsylvania to Kingston, New York. The canal was built to transport newly-discovered anthracite coal from Pennsylvania to the energy-starved metropolis of New York City. New York was desperate for a cheap, reliable source of fuel at this time (sound familiar?) due to trade restrictions and tariffs on British coal imports following the War of 1812. The canal's route down the Delaware River to Port Jervis, and then northeastward up the Rondout Valley to the Hudson River may seem like a roundabout way of getting there, but it allowed coal to be shipped economically on barges without having to traverse any intervening mountain ridges.

The canal was designed by Benjamin Wright, Chief Engineer from the newly-constructed Erie Canal, and it was a technological marvel. The canal, all dug and constructed essentially by hand, averaged 10 meters (33 feet) in width and crossed 137 bridges and 108 locks (one for each mile of its length, on average). The locks allowed for 325 meters (a little over 1,000 feet) of elevation change along its length. Barges, pulled by mules along the adjacent towpath, averaged 25 kilometers (16 miles) each day, and made a one-way trip in about a week.

The same year construction began on the canal, a blacksmith shop along the route in the tiny village of High Falls is credited with the discovery that the Rosendale member of the Rondout Formation could be made into natural cement. It didn't take long for commercial mining and production to develop in a narrow belt running northeast from High Falls to Kingston, and locally-produced "Rosendale Cement" was soon utilized in the construction of the D&H Canal. The D&H Canal Historical Society has preserved several of the canal locks, built of local Shawangunk conglomerate and Rondout Formation dolostone, in the area. The Society also operates a small museum in High Falls that is well worth visiting. While it's a sleepy little town today, you would not have recognized High Falls during its heyday when the canal was operational and rowdy canawlers filled the taverns on a Saturday night.

D&H Canal Lock 16 in High Falls

While in High Falls, be sure to visit the eponymous waterfall along the Rondout Creek. This stretch of property is owned by the Central Hudson Gas and Electric Corporation which operates a hydroelectric plant here and graciously allows public access to the property. Walking along the creek, one will see the resistant gray rock of the Rondout Formation capping the waterfall which is underlain by the greenish Binnewater Sandstone, red High Falls Shale, and white Shawangunk Conglomerate further downstream. A number of prominent faults and folds, visible on the opposite side of the creek, attest to a subsequent deformation event we'll discuss later. Note also the old stone ruins near the falls. This building was once a water-powered cement mill, as well as having earlier incarnations as a grist mill and fulling mill (where woolen cloth was felted by repeated blows from a mallet). A small limestone quarry and cement kilns are hidden on a nearby country road. It was a great boon to the builders of the D&H canal when natural cement was discovered locally (they would have had to ship cement from the Syracuse area otherwise) and the Rosendale cement industry benefited by being close to the route of the D&H canal, thus greatly reducing their shipping costs.

Natural cement was produced from a roughly north-south narrow belt of rock running between High Falls and Kingston in Ulster County.

The problem with mining in the local area is that the Rondout Formation is heavily folded and faulted and the dolostone had to be extracted either from surface quarries or underground mines depending upon the local geology. In some places, it crops out at the surface and is easy to quarry. Some of these old quarries are now filled with water forming pretty little lakes, such as the one behind the High Falls Motel on Route 213 just east of the village. In other places, the cement units dip deeply into the subsurface and require underground mining techniques. The underground mines can be a bit more difficult to locate than the surface quarries and some of them are surprisingly large. When standing in their cavernous rooms, it's hard to imagine that they were mostly excavated with sledge hammers, star drills (iron rods held by hand that other workers hit with the sledges), and black powder.

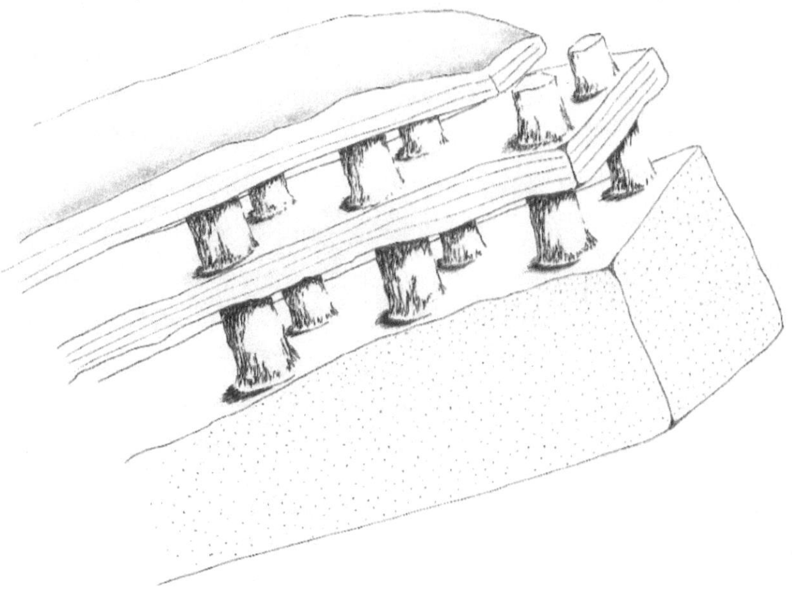

Room and pillar mining. In the Rosendale area, there is a "Lower Cement" layer and an "Upper Cement" layer. These are separated by a "Middle Ledge" limestone which was left in place.

One example of an underground cement mine is utilized by a company called Iron Mountain, located just outside the Village of Rosendale on Binnewater Road. Iron Mountain stores paper records and computer data for many large corporations in its climate-controlled underground buildings. These buildings are located in an old cement mine. I had a summer job there as a younger man and can verify that the mine was large enough to drive trucks into and contained large two-story buildings in its cavernous depths. Similarly-large mines, once used to commercially grow mushrooms, exist in the East Kingston area off Delaware Avenue. While some are quite extensive in area, local legends that maintain one can walk for miles in these underground mines are simply not true. This would only be possible if the Rondout Formation was horizontally bedded but, as we've already mentioned, it's extensively folded and faulted throughout the Hudson Valley.

Double-decker natural cement mine in Rosendale

Unfortunately, virtually all of these mines are on private property and off-limits for casual investigation – many are even dangerous since they were excavated long before OSHA regulations and federal safety inspections (and they've been unmaintained for over a century). There is, however, one excellent example of a mine that can be visited at the A.J. Snyder Estate off Route 213 in Rosendale. The property is owned by the Century House Historical Society and is home to the Widow Jane Mine, a classic example of a small room-and-pillar cement mine. The room is formed from the removal of rock and the pillars are left to hold up the mine's ceiling. The height of the ceiling represents the thickness of the Rosendale Dolostone member in this area. The floor of the mine is sloped because the Rondout Formation here is folded and dips down into the subsurface (some of the nearby mines in the area are slots following vertical beds downward). As with many of the natural cement mines in the area, the full extent of the mine is not really known because groundwater has filled the lower sections of the mine up to the level of the local water table. When the mine was operational, pumps ran day and night to keep it dry. Walking around to the side of the mine will allow you to view the overlying Glasco Limestone member which is chock full of stromatoporoid and chain coral fossils. Many nearby mines actually have two levels due to simultaneous mining of the lower Rosendale Dolostone and upper Whiteport Dolostone members (historically called the lower and upper cements by miners).

Once the rock is mined, natural cement is relatively easy to produce. After the rock was crushed into fist-sized pieces, layers of rock and coal were continuously fed into the top of a brick-lined kiln where the intense heat drives off carbon dioxide (CO_2) gas from the calcium-magnesium carbonate dolostone $(Ca,Mg)(CO_3)_2$ – a process called calcination. The resultant rock, removed from the bottom of the kiln, was lighter, more porous, and easily crushed into a powder. The powder was a natural cement, just add water and it essentially reverted right back into solid rock. Especially useful was the fact that this was hydraulic cement, meaning it would harden underwater. The cement powder was placed into wooden barrels (bags were later used), transported to D&H Canal barges, and shipped to the Hudson River and from there to the outside world.

Rosendale cement kilns

Much of the Rosendale natural cement wound up in New York City which was undergoing a building boom throughout the 19th century. Some of the more famous structures constructed from this cement were the Brooklyn Bridge and the pedestal for the Statue of Liberty. Some wound up much further afield since, at one time, the Rosendale District provided roughly half of the natural cement used in the United States. Rosendale cement was even used to construct Fort Jefferson in the Dry Tortugas, 70 miles west of Key West, Florida. By 1900, the Rosendale district was producing nearly ten million barrels of cement each year and employing thousands of people. Unfortunately for Rosendale, natural cement fell out of favor in the early 1900s with the commercial success of Portland cement. Portland cement can be manufactured from many different types of limestones by controlling its chemical make-up through selected additives. This yields a product with a uniform composition and predictable physical properties (important in the modern building industry). The last big project partially constructed with Rosendale cement was the New York State Thruway in 1954 and by 1970 the last company had shut down. Interestingly enough, there is a renewed demand for natural cement in historical renovations and very small-scale production has resumed in the area.

The deposition of the Rondout Formation basically closes down the Silurian Period (although the exact moment of the Silurian-Devonian boundary probably lies somewhere within the overlying rock formation). The subsequent Devonian Period was an interesting and eventful time in Earth history. There was a rich diversity of life in the seas, now including vertebrate fish, and by the end of the Devonian there was an assemblage of plant life, insects, and amphibians on land as well. While plants and animals had starting coming onto land earlier, they were only beginning steps. By the Devonian, we start to see the fossil remains of entire forests – full-blown ecosystems that we'll discuss in Chapter 9.

The big story in the Hudson Valley for the Early Devonian Period is the formation of the Helderberg Sea. To introduce you to the rocks formed in this sea, I'd like to take you on a guided tour of an outcrop found along Route 199 just west of the Kingston-Rhinecliff Bridge in Ulster County. This outcrop is relatively safe to visit if you pull well off the busy road. The rocks exposed here are very instructive in showing how conditions changed over time in the Helderberg Sea. I take my geology students to this outcrop, and others nearby, every semester. Let's first take a brief digression to discuss limestone in a little more detail than we have thus far.

In introductory geology classes, we teach students how to identify sedimentary rocks. One common example is limestone, a carbonate rock composed of the mineral calcite ($CaCO_3$) which we've already discussed. It's important to understand, however, that there are many different types of limestones. As a matter of fact, geologists have developed several classification schemes to distinguish between the various types. So, while all limestones are composed of calcite, not all limestones are the same in detail. Limestones may be composed of pure, fine-grained carbonate mud (called micrite or calcilutite depending on the classification system used), coarsely-crystalline calcite (sparite or crystalline limestone), calcite with some quartz sand mixed in (arenaceous limestone or calcarenite), large amounts of fossil material (fossiliferous or skeletal limestone), small spheres of calcite (oolitic limestone), and many other things. Because there are different classification schemes, the exact same piece of limestone may actually be called different names by two or more separate geologists which can obviously be confusing for those not up on the current jargon.

Why are there different types of limestone? Well, it turns out that they form in different environments. Virtually all limestones form in warm, shallow seas. A few are lake bed deposits, but those don't exist here in the Hudson Valley so we won't worry about them. If the shallow sea has a lot of coral reefs and other calcareous invertebrates, a fossiliferous limestone might form. A quiet environment with fine-grained carbonate mud might be characterized by a micrite. An influx of sand mixing with the carbonate mud would form a calcarenite. In addition to the type of limestone, sedimentary structures are also important to understand. Mud cracks preserved in the limestone indicate its position on a tidal flat while ripple marks indicate a shallow marine setting. Fossils are also key indicators of ancient environments. Paleontologists can often make rough estimates of the depth of seawater by examining the fossil assemblages in the rock. Some organisms prefer shallow water with a lot of wave energy. Others prefer deeper, quieter water. If fossils are lacking, the environment may have been hypersaline or otherwise hostile to life.

Let's now examine this outcrop in more detail. It's located north of the city of Kingston, on the north side of Route 199 just west of the Route 32 overpass. We'll start at the east end of the outcrop, closest to the Hudson River, and work our way westward. The first thing you might notice is that these rocks are tilted such that they're sloping downward toward the west. Geologists would say that they're dipping to the west. Assuming that they were simply tilted from horizontal (not upside-down), it's obvious through the principle of superposition that the rocks will be older to the east. So as you walk from east to west, you're moving upward in the stratigraphic sequence. These tilted rocks are actually part of a larger-scale fold as we'll soon see.

The first limestone unit seen on the eastern end of this outcrop is called the Manlius Formation. It's named after a small town near Syracuse giving some idea as to its widespread extent in New York State. The Manlius is a fine-grained, lighter gray limestone (a micrite) and has very characteristic thin bedding in places. These very thin laminations, often called "ribbon beds," along with the fine carbonate mud composition and occasional mud cracks lead geologists to believe that this unit was deposited in a quiet, lagoonal type environment. This is similar to the underlying Rondout Formation we've already discussed. The Rondout is not exposed right at this outcrop, but can be seen below the Manlius in the nearby woods.

Tilted rock layers on Route 199 in Kingston

There are a few fossils which can be found within the Manlius. One of my personal favorites is *Tentaculites gyracanthus*. *Tentaculites* fossils look like ribbed cones and are only a few millimeters in length. I've personally collected slabs of Manlius limestone from other localities covered with thousands of them — they were quite numerous in this environment. The interesting thing about *Tentaculites* is that no one is quite sure exactly what it is since the soft tissues were not preserved. While most paleontologists classify them into the gastropod group (marine snails), others have argued that they may be a type of annelid (segmented worm).

Another group of fossils common within the Manlius Formation are ostracods (*Leperditia alta*). Ostracods are a type of crustacean sometimes called seed shrimp due to their small size, a few millimeters in length, and their seed-like carapace. These ostracods probably lived within the upper layer of the bottom sediments making them relatively easy to find on Manlius bedding planes with careful examination.

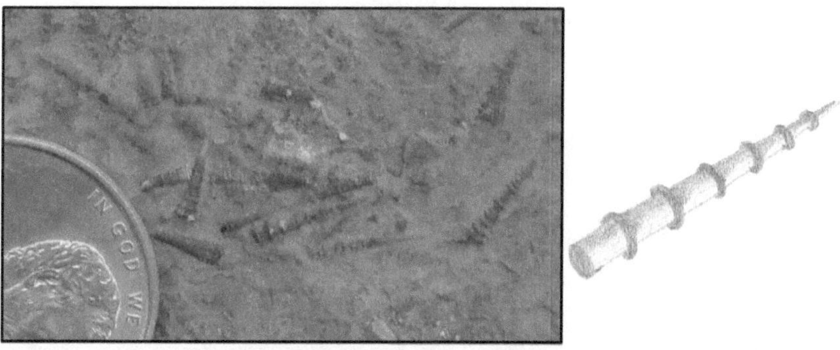

Photo and reconstruction of Tentaculites gyracanthus (penny for scale)

The upper part of the Manlius is a bit different from the lower part and becomes coarser-grained with irregular bedding. The irregular bedding results from layers of stromatoporoids (calcareous sponges). These stromatoporoids formed patchy reefs that were apparently affected by wave action. Water levels were increasing and the environment was changing from a shallow lagoonal setting to a higher-energy, reef environment.

As you slowly work your way along the outcrop, you'll soon notice that the limestone color has changed. While the Manlius Formation is typically light gray, the next unit changes to a darker gray. Also, unlike most of the Manlius, this next limestone has a much coarser texture with visible grains and thicker bedding. This limestone is called the Coeymans (pronounced "cwee-mans") Formation. The Coeymans was named after a small town on the Hudson River in southern Albany County.

Stratigraphic contact between the Manlius and Coeymans Formations

The Coeymans Limestone represents a better development of the reef system beginning in the upper part of the Manlius Formation. Tabulate corals are abundant as are crinoids and brachiopods. Crinoids and brachiopods were both discussed in the Prelude chapter so we won't say much more about them here other than mentioning the Coeymans' distinctive index fossil, the brachiopod *Gypidula coeymanensis*. Look around this area of the outcrop and you'll find thin layers of broken up *Gypidula* shells, some of which might represent ancient Devonian tropical storm deposits.

Fossil "hash" of Gypidula coeymanensis possibly representing a storm deposit

If you continue westward along the outcrop, you'll soon arrive at the third limestone unit – the Kalkberg Formation. Kalkberg is Dutch for "lime mountain" where lime, of course, refers to calcium carbonate, not the fruit! The Kalkberg is easy to recognize because it has nodules and stringers (elongate masses) of black chert embedded within a medium-grained limestone. Chert is cryptocrystalline quartz (quartz where the crystals are so fine they're difficult to see even under magnification) and commonly forms within limestones through a variety of mechanisms. The chert within the Kalkberg may well have been derived from the alteration of volcanic ash since the Kalkberg also contains a number of bentonite clay beds. As with the earlier Ordovician shales, these bentonite clays represent volcanic ash-falls into the Helderberg Sea. These ash layers date to a little over 400 million years ago which places them in the Early Devonian Period. We'll discuss the volcanoes that produced these ash layers in the next chapter.

The Kalkberg is richly fossiliferous with brachiopods, corals, crinoids, trilobites, bryozoans, and other bottom-dwelling (benthic) marine invertebrates characteristic of clear ocean water. Bryozoans are minute, colonial organisms related to brachiopods. Both groups use a filter-feeding organ called a lophophore to strain nutrients out of the seawater. Bryozoans don't build calcareous hinged shells; they do, however, develop coral-like structures in which to live. There are several groups of bryozoans; some forming crusts on other objects such as corals or rocks (waving tentacles protruding from encrusting bryozoans led to their common name of "moss animals"), others develop branching, tree-like stalks, and yet others form lacy fans. Fragments of these calcareous bryozoan structures are commonly preserved as fossils within limestones like the Kalkberg.

Bryozoan fragments in a Helderberg Group limestone

The final rock unit exposed at this outcrop is the New Scotland Formation, named after a town in western Albany County. As you approach the western end of the outcrop, you'll see the chert start to disappear and the rock takes on a new appearance with buff-brown layers alternating with medium-gray layers. The brown layers represent beds with a little more iron oxide mixed in with the carbonate sediments – the rock essentially rusts as it weathers. The New Scotland Formation limestone contains a fair amount of clay and in places is actually classified as a calcareous mudstone.

From east to west along the outcrop, the limestones formed in increasingly deeper waters and reached their maximum depth with the New Scotland Formation. The bottom sediments were not affected by ordinary wave action although large storms would likely have disturbed the seabed. The New Scotland also contains abundant fossils; over 300 species have been identified from this unit. If you walk all the way to the western end of this outcrop, and then climb onto the bedding plane, you will be standing on the Devonian Period seafloor. Looking down will reveal a limestone surface literally carpeted with brachiopod fossils underfoot. It's a good place to sit and reflect on how drastically our part of the world has changed over time.

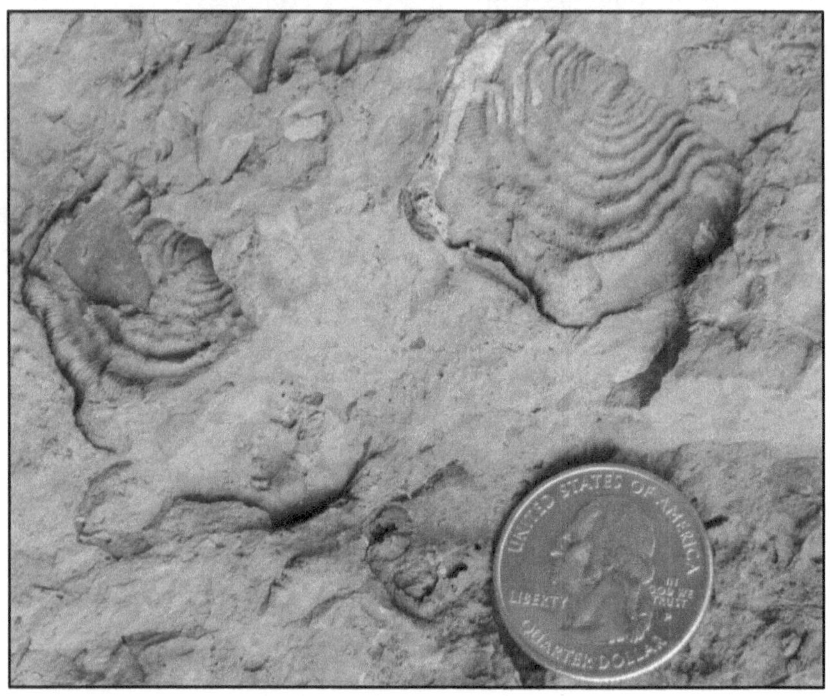

Fossils of the brachiopod Leptaena rhomboidalis in the New Scotland Limestone

As we've walked our way up the length of this small outcrop, we've seen how changing conditions over time have led to four distinct limestone units (Manlius, Coeymans, Kalkberg, and New Scotland) being deposited here. How much time do these rocks represent? It's impossible to provide an exact number. Limestones can't be radiometrically dated, but we can estimate a ballpark figure. I've paced out the length of this outcrop at 120 meters (almost 400 feet) and the beds are dipping about 30° from horizontal. Simple trigonometry (120 meters x sin30°) gives us a cumulative sedimentary rock thickness of about 60 meters. Studies have shown that typical carbonate sedimentation rates are around 5 centimeters of sediment every 1,000 years (this is a very conservative estimate; rates can be much higher under favorable environmental conditions). Given this average rate, this particular outcrop apparently represents only 1,200,000 years of sedimentation (this doesn't include time for the sediment to later be buried and lithified into limestone). While 1.2 million years may sound like a long time, at least to non-geologists, the fact is that the Early Devonian Period lasted for 18.5 million years. The deposition of these limestones would have taken only 6.5% of the total time available to deposit all of the Helderberg Sea sediments.

If you would like to see additional examples of limestones, walk or drive a short distance west to the next large outcrop – a spectacular folded arch of limestone. This outcrop is actually pictured in a number of geology textbooks as a classic example of a type of fold called an anticline. In the center of this fold is a coarsely-crystalline, light-colored limestone called the Becraft Formation. The Becraft is composed almost entirely of crinoid stem fragments in places (crinoids were also discussed in the Prelude). This fragmentation indicates a high-energy environment and therefore a lowering of sea level from the previously-deposited New Scotland Formation. During this time, dense "meadows" of crinoids would have covered the shallow seafloor.

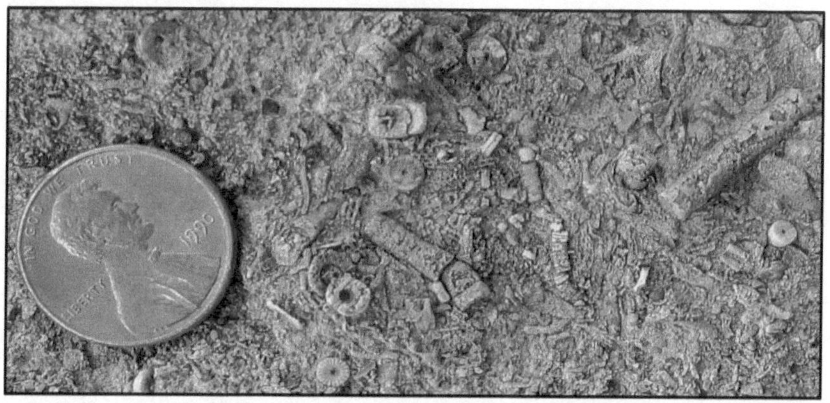

Crinoid stem fragments in a Helderberg Group limestone

Looking at the fold from across the road, you can easily see there are two darker units above the lighter-colored Becraft Formation. These are called the Alsen and Port Ewen Formations. The Alsen and Port Ewen are analogous to the Kalkberg and New Scotland, respectively and represent a second deepening of sea water here forming argillaceous (clay-rich) limestones which are shaly in places.

Driving a bit further west on Route 199 will bring you to a third outcrop, this time showing a downward-sagging fold. Geologists call these folds synclines (remember them as anticline/arch and syncline/sag). This outcrop contains the younger Esopus Shale and

Anticline along Route 199 in Kingston

Zoophycos cauda-galli feeding traces in the Esopus Shale

Schoharie Formation which lie stratigraphically above the rock units seen at the anticline. A very interesting fossil can be seen here in the Esopus Shale. Climbing onto a prominent bedding plane in the middle of the outcrop will reveal a smooth, glacially polished surface with funny-looking swirls scattered throughout. These swirls are actually the preserved feeding traces of a marine worm called *Zoophycos cauda-galli*.

Besides the mud-eating worms, the Esopus Shale is otherwise not very fossiliferous at this location. These deeper waters may have been somewhat anoxic (oxygen-poor) at the time.

Another superb place to view the Helderberg Group of limestones is 100 kilometers (60 miles) north of Kingston at Thacher State Park. Located west of Albany, Thacher Park contains the Helderberg Escarpment, a line of cliffs composed of the Manlius and Coeymans Formations (other Helderberg Group units are found west of the escarpment). The park contains a wonderful trail that takes you from the top of the cliff down to the bottom where you'll walk behind a waterfall (Minelot Falls) and pass several natural springs before climbing back up to the top of the escarpment. The path is called Indian Ladder Trail and is located where an actual ladder was once located to assist in scaling the cliffs here. This trail was once used by Mohawk Indians who traveled between the Schoharie Valley and the frontier trading post at Albany.

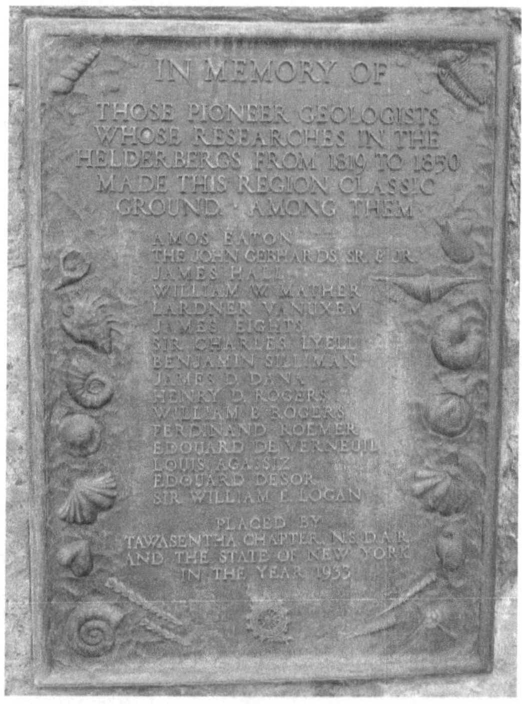

Plaque on Indian Ladder Trail

This is geologically hallowed ground. A memorial plaque at the start of the trail honors many of the famous geologists of the early 19th century who spent time here studying these rocks. The plaque also has some relief sculptures of a few of the fossil organisms found in this area. Generations of geologists have tromped through here to learn the stratigraphy of the Helderberg Group. Even today, vans full of geology students from all over the Northeast will visit these classic exposures. It's also a beautiful place to spend a summer afternoon with views across the Hudson Valley to the distant Green Mountains of Vermont. When geologists stand on these cliffs, they don't just see the low, rolling Taconic hills to the east but rather, in their mind's eye, the towering snow-capped peaks of the once-mighty Taconic Mountains.

Following the deposition of the Helderberg Group of rocks, the sea once again regressed from New York creating an unconformity. A new transgression deposited another series of sedimentary rocks on top of this erosional surface. These are called the Tristates Group and include the Esopus Shale and Schoharie Formation discussed earlier. They're also found throughout the Hudson Valley but we'll forego their detailed descriptions other than to note they are mostly carbonate units similar to the underlying Helderberg Group. This was followed by another regression and the formation of a regional unconformity. We're now in the Middle Devonian Period and yet another marine transgression deposited the richly-fossiliferous Onondaga Formation limestones.

Why all of these marine regressions and transgressions? The regressions are easy to explain. In the Early Devonian Period, the Tippecanoe transgression was waning and sea levels were dropping all over the world. The transgressions are a bit more problematic. The area covered by the Helderberg Sea and subsequent transgressions was relatively flat with little topographic relief. It would not have taken much of a change in sea level to flood this area. There's no evidence of continental glaciation during this time so that probably wasn't responsible for the sea level fluctuations but there was a lot of plate tectonic activity which may well have affected local sea levels. This plate tectonic activity will be explained in the next chapter where we'll see how these rocks became so deformed.

The Onondaga Formation is worth discussing since it's found throughout the mid-Hudson Valley, as well as further afield (it extends past Buffalo to as far as Detroit to the west), and is very distinctive. The Onondaga is subdivided into four members (Edgecliff, Nedrow, Moorehouse, and Seneca), which are generally analogous to the older Coeymans, Kalkberg, and New Scotland Formations in terms of their depositional environments in a carbonate sea. This rock unit is also extremely fossiliferous containing large numbers of coral reefs within its strata. Abundant chert within the Onondaga, especially in the Moorehouse Member, cause it to weather with a distinctive "knobby" appearance due to the different rates of weathering between the relatively soft limestone and much harder chert nodules.

After looking at the Helderberg Group rocks along Route 199 in Kingston, drive down Route 9W south to see exposures of the Onondaga Formation along the east side of the road and behind many of th parking lots and businesses along this highly-developed strip. The knobby appearance of this white limestone is unmistakable. Forsyth Park, on the south side of Kingston off County Route 1 (Lucas Avenue) also has outcrops of Onondaga limestone you can walk on and examine at your leisure.

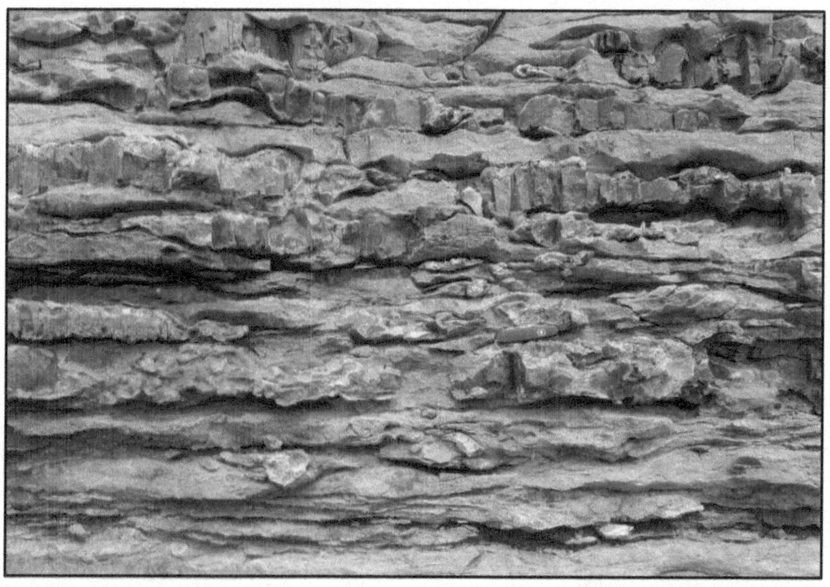

Weathering of chert-rich Onondaga Limestone (Swiss army knife for scale)

Near the top of the Onondaga Formation, there are a number of famous bentonite layers called the Tioga ash beds. These are a series of volcanic ash layers deposited by several extremely large eruptions in the Middle Devonian Period radiometrically dating to around 390 million years ago. The Onondaga limestones represent the last record of Devonian seas within the Hudson Valley and this volcanic ash is foreshadowing another mountain-building event we'll discuss shortly.

One important group of features found in many Devonian Period limestones in the Hudson Valley has to do with the dissolution of the limestone by acidic groundwater. This dissolution is a relatively modern feature, mostly occurring since the last Ice Age, but will be mentioned here since it's well developed in the Manlius and Onondaga Formations.

The most impressive of these dissolution features are underground cave systems.

It's outside of the Hudson Valley, but Schoharie County, west of Albany, is famous for its caves. Many of them are developed in two different limestone beds – the Manlius Formation and the Onondaga Formation. There are also numerous sinkholes, disappearing streams, and springs indicating the extent of limestone dissolution in the area. These types of features are characteristic of what geologists call karst topography, a term which comes from the Krs Region of Slovenia (a place apparently lacking in vowels). Howe Caverns is a commercially-developed cave that is worth visiting and developed within the Manlius Limestone. There's a large limestone quarry next to Howe Caverns where this limestone was mined for use in Portland cement. As a matter of fact, you would never know by looking at it but there's also an old, abandoned cement mine beneath the quarry in the underlying Rondout Formation. Many of the undeveloped caves in this area are administered by cavers and can be visited by joining caving organizations like the Albany area Hudson-Helderberg Grotto.

There are also a few small caves (not to be confused with man-made cement mines) in the Rondout Valley from Kingston southwestward into the Rondout Valley. One of the more interesting is Pompey's Cave near High Falls which also exists within the Manlius Limestone. Pompey was reputedly the name of an African-American farmhand who discovered the cave. An unverified, and rather unlikely, local story is that this cave was used as part of the Underground Railroad to temporarily house runaway slaves. Pompey's Cave is a several hundred meter long tunnel, tall enough to walk through, in which a permanent stream is flowing on its way to the nearby Rondout Creek. The stream once flowed on the surface, where a dry rocky streambed now exists, and the cave is accessible through a small hole in the bedrock and 3 meter (10 foot) high ladder. Pompey's cave is easy to explore since it's a tunnel that dead-ends in either direction, but it can be dangerous during high-water conditions. It can completely fill during a heavy rainstorm with the water returning to surface flow in the normally-dry stream bed.

On the surface, walking upstream from Pompey's Cave will soon bring you to a swampy area sitting on top of fractured limestone bedrock. This is where water seeps downward into the cave system. Walking downstream from the cave opening will bring you to a natural spring, on private property, where the stream resurges back onto the streambed and continues as normal surface flow to the Rondout. In the nearby woods, the limestone bedrock exhibits other areas of solutionally widened joints (some large enough to crawl into) and areas where the ground has subsided due to collapse of underground voids. These subsidence structures are called sinkholes.

This small cave system illustrates several important principles about groundwater flow in karst areas (karst hydrology). First, groundwater flowing through soluble limestones like the Manlius Formation will often dissolve solution joints and even form larger cave passages. On the surface, a number of sinkholes may be found where cavern roofs collapse. Disappearing streams and resurgent springs may be common. These are all features found in this area of the Hudson Valley. Secondly, water flows very quickly through areas of karst development. Water-tracing dye tossed into the swampy area upstream of the entrance to Pompey's Cave will soon be seen emerging from the resurgent spring downstream from the opening. This is not how water normally flows through bedrock. When water oozes through the microscopic pore spaces in rock, it only moves a few centimeters a day. When water moves through a karst system, however, it's essentially flowing through an open pipe and can travel a long distance in only a few minutes. When water oozes slowly through pore spaces in rock, endemic bacteria naturally clean the water over time. In karst areas, this cannot occur and they are therefore susceptible to contamination.

Entrance to Pompey's Cave in a dry stream bed

The environmental sensitivity of karst areas is a critically important issue in some areas of the Hudson Valley where ever-increasing development and urban sprawl are occurring in areas underlain by

Devonian limestones. In rural areas, people obtain their groundwater from private wells and dispose of their wastes with septic systems. In such cases, it's essential to pay attention to the geology beneath your home. We'll return to some of these environmental issues at the end of the book.

Back to the Devonian Period. As marine invertebrates happily lived within the clear, sunlit waters of the Helderberg Sea, we can begin to cue the ominous music. Something's approaching their idyllic home. Something large and menacing. What's approaching is a microcontinent called Avalonia. The inexorable movement of tectonic plates brings it closer to Laurentia each year. Soon, the suspension-feeding invertebrates in the clear seas start to notice some mud in the water; mud which clogs their feeding organs and gradually replaces the carbonate sediments of the seafloor with clastic sediments. Limestones turn to muddy limestones and then to shales. Organisms die as the sea relentlessly dries up. The collision occurs; a massive mountain-building event of a magnitude not seen since the formation of the Grenville Mountains almost a billion years earlier. This event, the Acadian Orogeny, is the subject of our next chapter.

8
Avalonia

We felt ourselves necessarily carried back to the time when the schistus on which we stood was yet at the bottom of the sea, and when the sandstone before us was only beginning to be deposited, in the shape of sand or mud, from the waters of a superincumbent ocean. An epoch still more remote presented itself, when even the most ancient of these rocks, instead of standing upright in vertical beds lay in horizontal planes at the bottom of the sea and was not yet disturbed by that immeasurable force which has burst asunder the solid pavement of the globe. Revolutions still more remote appeared in the distance of this extraordinary perspective. The mind seemed to grow giddy by looking so far into the abyss of time.

John Playfair,
Illustrations of the Huttonian Theory of the Earth

John Playfair, in the above quotation, was discussing a 1788 field trip taken with Edinburgh geologist James Hutton (1726-1797) to examine rocks along the southwestern coast of Scotland. One particular outcrop they examined, at a place called Siccar Point, is now a geological icon pictured in virtually all modern geology textbooks. While historically important, and quite scenic, Siccar Point is not unique as far as geological outcrops go; we actually have a very similar version of it here in the Hudson Valley. This location was, however, where geologists for the first time really thought about what such outcrops can tell us about the history of the Earth.

At Siccar Point, Upper Devonian red sandstones are sitting in angular unconformity atop Lower Silurian graywackes (the "schistus" Playfair speaks of above). The angular unconformity is itself tilted as well. Geologists today know that there are about 80 million years of time missing in this unconformity but this was not known in Hutton's day. Hutton and his party did recognize, however, the vast amount of time represented here.

Hutton's Unconformity at Siccar Point

The formation of a tilted angular unconformity between two sedimentary rock units reflects a long sequence of events. The Lower Silurian graywackes had to first be deposited in deep marine waters (they're turbidites that formed near a subduction trench). The sediments needed to be buried and lithified into sedimentary rock. Then the sedimentary rock had to be folded during a mountain-building event. Since the folding occurred at depth, the graywackes needed time to be eroded and brought to the surface. Now we're in the Upper Devonian Period. Once at the surface, the graywackes accumulated sediments on top of their eroded surfaces. The whole package of rock and sediments then needed to be buried again to lithify the new sediments into red sandstones. Another mountain-building event occurred to fold everything once again since the overlying sandstones aren't horizontal either. Finally, erosion occurred to expose these rocks to modern-day weathering by waves along the coast of the North Sea. Every step in this process takes vast amounts of time.

The systematic study of outcrops, such as that at Siccar Point, sounded the death knell for the traditional Biblical age of the Earth. There's no conceivable way, short of arbitrary miraculous interventions, that such a sequence of events could occur in only a few thousand years. While Hutton knew nothing of plate tectonics or radiometric dating, he did know that this outcrop represented an almost inconceivable span of time. Enough time to make the mind "grow giddy by looking so far into the abyss of time," in the words of Hutton's colleague, John Playfair.

While it would be nice to travel to Scotland and make a pilgrimage to Siccar Point, it's not really necessary. We have our own version right here in the mid-Hudson Valley called the Taconic Unconformity. Along Route 23, just west of Catskill, is a remarkable sequence of folded and faulted rocks. If you are travelling westward from Route 9W, take the exit for Route 23B and carefully pull over next to the outcrop along the exit ramp. At the east end of this outcrop, you'll see shales and interbedded sandstones with nearly vertical bedding. These belong to the Ordovician-age Austin Glen Formation of the Normanskill Group and, as previously discussed, formed between the Taconic Island Arc and Laurentia. These shale and graywacke turbidites can easily be traced eastward to the Hudson River visible in exposures where Route 23 crosses the Rip Van Winkle Bridge.

Taconic angular unconformity off Route 23 near Catskill

To the left of these shales is a tilted contact overlain by an orange-brown dolostone – the Late Silurian Rondout Formation. This rock unit is much thinner here than it is to the south in Rosendale but just as recognizable. It quickly grades upward into our old friends, the gray limestones of the Helderberg Group (Manlius, Coeymans, Kalkberg, and the New Scotland Formation at the west end of the outcrop). Note that all of these overlying carbonate rocks are dipping downward toward the west (but are not near vertical like the underlying Ordovician shales). This contact, between the Austin Glen shales and Rondout dolostone is a classic angular unconformity that is very similar to that seen in far-off Scotland.

The origin of this angular unconformity in the Hudson Valley is almost exactly analogous to how Siccar Point formed. The first step was to deposit sediments on the deep seafloor during the Ordovician Period. The shales and sandstones we talked about earlier when we discussed the approach of the Taconic island arc. These sediments accumulated in horizontal layers according to Steno's principles. They didn't form overnight either. It takes time for sediments to accumulate and it takes time for sediments to become buried deeply enough such that they're lithified into sedimentary rock. Then, sometime after these sediments became sedimentary rock, they were folded. We already know that this was from the collision of the Taconic island arc with Laurentia – the Taconic Orogeny. Then, after folding, these rocks were eroded down to form a flat surface again. Or, more strictly speaking, they were folded deep underground and mountains of rock above them were eventually eroded away to expose them at the Earth's surface. The area became low enough to be a seafloor again. That's a lot of erosion and about 60 million years of time is now missing here. The seafloor was under the Devonian Helderberg Sea where carbonate sediments were deposited on this ancient erosional surface. They were, of course, deposited in horizontal layers resulting in an angular contact with the tilted Ordovician sedimentary rocks beneath them. Everything was again buried deep enough to lithify these carbonate sediments into dolostone and limestone. Once again, the whole area was folded. We know that because the angular unconformity is itself folded. The second mountain building event affecting these rocks is a bit more problematic to identify but there are two possible candidates.

The first possibility is that these rocks were folded by the Acadian Orogeny, the subject of this chapter. This was a mountain-building event primarily caused by the collision of Laurentia with a microcontinent called Avalonia starting in the Middle Devonian Period. The second possibility is that these rocks were instead folded by a later mountain-building event called the Alleghenian Orogeny and caused by the collision of Africa with Laurentia to form the supercontinent of Pangaea near the end of the Paleozoic Era. We'll discuss this orogeny in more detail in a later chapter.

I happen to believe that these rocks in Catskill were folded by the Acadian Orogeny but there are a number of geologists who would disagree with this interpretation. The problem is that there is no easy way to determine exactly when sedimentary rocks were folded. We'll come back to this later in the chapter and discuss the pros and cons for each of these two possibilities.

Up to this point, we've been discussing the appearance of rock layers in a very general sense. Rock strata may be horizontal, tilted, or even near vertical in their orientation. When rocks are tilted, we can speak about which way they're dipping (the tilted Taconic Unconformity near Routes 23 and 23B is dipping toward the west, for example). In science, however, we have to be a bit more precise about such things and geologists have therefore devised a method to quantify the orientation, or strike and dip, of rock layers (it works for other planar features such as faults, dikes, or joints as well). Geology students learn how to measure the strike and dip of rocks when they study structural geology – the branch of geology dealing with the deformation of rocks.

Dip is easier to understand so I'll explain that first. The dip of a rock layer is simply its angle from horizontal. If a rock layer is horizontal, its dip angle is 00°. If a rock layer is vertical, its dip angle is 90°. All other dip angles fall between these two extremes. You can also specify the direction in which a rock layer is dipping. If you specify a dip as 30° to the southwest, for example, this would mean that if you dropped a marble on the rock layer, it would roll down a 30° incline toward the southwest.

Strike is a bit more conceptually difficult. Imagine a dipping rock layer. Now imagine flood waters rising around this rock layer to half cover it with water. The water surface will intersect the dipping rock layer along a horizontal line (the surface of quiet water is always perfectly horizontal). We can call that the strike line. Strike lines are horizontal lines on a dipping rock layer and are also exactly perpendicular (at right angles) to the rock's dip direction. The strike of the rock layer is the compass direction, or azimuth, of that line. The problem is that lines have two ends. Since any strike line is pointing in two opposite directions, which end do we specify as the strike direction? It's customary to imagine standing on the outcrop and facing the end of the line such that the rock is dipping to our right and specify that as its strike direction. For example, if the rock strata are dipping to the west, the strike line will have two directions, north and south. If we stand on the outcrop so the dip direction (west) is to our right, we'll be facing south and we'll then specify this as the strike of the rock layer. This is called the right-hand rule.

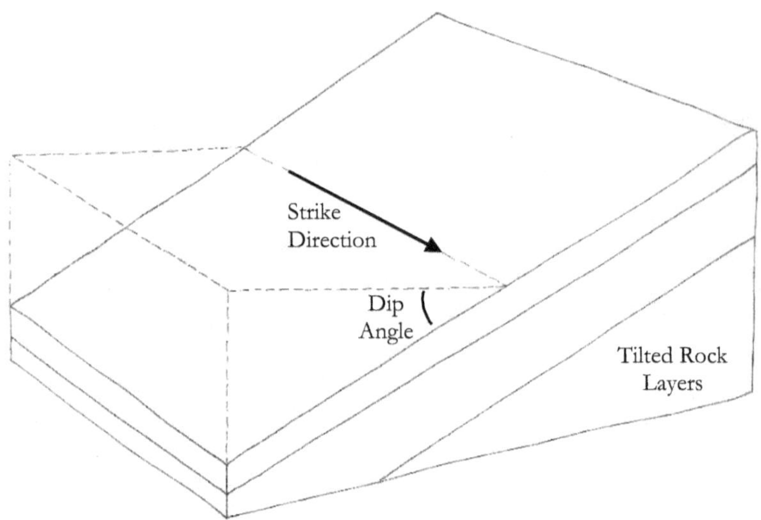

Strike and dip defined on a tilted rock layer

When specifying strike, geologists use the azimuth system rather than a simple compass direction in order to be a bit more precise. We traditionally divide up a circle into 360°, a cultural artifact from the ancient Babylonians. If we imagine a compass face, north is then assigned a value of 000°, east is 090°, south is 180°, west is 270°, and we return to north at 360° resetting the value to 000°. So, rather than saying a rock layer has a strike of southwest, we instead say it has a strike of 225°.

You might have noticed that I specified the dip of a horizontal bed as 00° and a north-trending strike line as 000°. This is a habit I try to encourage in my students since it allows us to be unambiguous when recording the strike and dip of a rock structure. If I see a strike and dip recorded as 005°, 10°, I know instantly which number is the strike and which is the dip. This is not necessarily the case if the student only records 5°, 10° for the rock's orientation. Strike and dip are recorded by special geological compasses called Bruntons® (after the company which manufactures them. These compasses have built-in levels allowing us to hold the compass horizontal when measuring strike lines and to easily measure the dip angle of a rock layer from horizontal.

Don't worry if this explanation is a bit confusing, students are often confused as well until they use a compass and measure a few dozen orientations in the field. I did want to take the time to explain the concept, however, because it gives an important insight into how structural geologists are able to infer geologic structures they can't directly see or measure. This is best explained by a real-life example.

Measuring strike and dip with a Brunton compass

Let's return to our rock outcrop along the Route 23B exit ramp from Route 23 west. Above the Taconic Unconformity which we've already discussed, there is the sequence of Helderberg Group limestones dipping to the west. Walking along the outcrop, you should be able to recognize the ribbon beds of the light-gray, fine-grained Manlius limestone, find some Gypidula fossils within the dark-gray, coarse-grained Coeymans limestone, identify the chert layers within the Kalkberg limestone, and note the prominent gray-brown striping of the New Scotland Formation closest to the bottom of the ramp. With a Brunton compass, I can easily measure the strike and dip of these beds as 190°, 40°. Since the beds are striking roughly south (10° west of south to be exact), the right-hand rule tells us they're dipping to the west (they're dipping away from the Hudson River behind us and toward the Catskill Mountains in the distance). Note how this implies that these rocks are dipping down under Route 23B at the bottom of the hill.

Let's cross Route 23B and get on the entrance ramp back toward Route 23 west. There is another long rock outcrop we can examine on this side of the road. From east to west (bottom of the ramp to the top), we'll see the following rock formations – New Scotland, Kalkberg, Coeymans, and Manlius. The difference now is that these rocks have a strike and dip of 345°, 30°. In other words, they're dipping approximately to the east, just the opposite of the exact same limestone formations across the street. What's going on here?

Rock layers on the east side of Route 23B. These layers are dipping downward to the west (left)

Rock layers on the west side of Route 23B. These layers are dipping downward to the east (right). Between these two rock outcrops, under Route 23B, is a fold

What we have are layers of sedimentary rocks, initially deposited in horizontal beds, now forming a downward-sagging fold beneath Route 23B. Geologists refer to these types of folds as synclines (remember syncline/sag?). We can't directly see the syncline, but the orientation of these rocks, along with our superior rock identification skills, tells us it must be there. This is the power of structural measurements, they allow us to "see" underground. Now that we know the orientation of the syncline, we can also infer that these rocks were compressed laterally to warp them downward into this type of fold.

While this is a very simple example of an obvious geologic structure, geologists apply these same powerful techniques to study regions of deformed rocks where the structures are much more complicated and larger in scale. The purpose of this is not only to describe the three-dimensional geometry of the rocks but to then be able to say something about how they got that way since their present-day orientation is a reflection of their ancient geologic history. By identifying our syncline, we now know that this area experienced east-west compressive stresses at some point after the Helderberg Group was deposited.

Before we leave this outcrop, there's one more feature here that's worth examining. As already mentioned, walking westward along the outcrop will bring you from the younger New Scotland Formation down into the older Manlius Limestone. Continuing along the outcrop, however, will reveal something extraordinary. Below the Manlius Limestone is not the Rondout Formation, as we would expect, but rather a reappearance of the chert-rich Kalkberg Limestone. Walking up and down the outcrop, we can confirm that we are seeing the Kalkberg Limestone stratigraphically above the Manlius and Coeymans Limestones as well as stratigraphically below them. How can this be true? Have we made some major mistake in our identification of the rock units?

Since we would expect the Kalkberg to lie above the Manlius, as it does everywhere else, let's look for the problem below the Manlius. If you examine the contact between the Manlius and underlying Kalkberg Limestone, you will see a several centimeter thick band of white calcite. What's more, this calcite layer is not parallel to the sedimentary bedding in the limestone – it cuts across the bedding planes. It turns out that we're looking at a fault; a fracture in the rock along which there was relative movement. As these rock layers were folding, the stresses were too much for the rock and it fractured allowing the Manlius Formation to slide up a fault ramp such that it's now lying atop the younger Kalkberg Formation. Water circulating through this fault precipitated the mineral calcite during deformation. This type of fault, emplacing older rocks on top of younger rocks, is called a thrust fault. If you look carefully along the fault, you will also find scratches in the rock, called fault striations, parallel to the direction of fault movement.

White layer in the middle of the image running from tree diagonally downward to the right is a mass of calcite crystals inside of a thrust fault. Note the slightly different orientations of the rock layers above and below the fault

These two types of structures, folds and thrust faults, are characteristic features in areas of deformed rock called fold-thrust belts. Fold-thrust belts form due to compression of layered sedimentary rocks during mountain building and these outcrops along Route 23 are a classic small-scale example. University of Illinois structural geologist Stephen Marshak has extensively mapped this area and refers to it as the Hudson-Valley fold-thrust belt (abbreviated HVB). While we're describing the HVB in the Catskill area, it essentially runs in a narrow belt from Albany southwards to Kingston (and perhaps into the Rondout Valley as well, depending upon how you define it).

As we've seen before, rock placed under stress can basically behave in two ways – it can fold (plastic deformation) or it can fault (brittle deformation). Folds, however, can only occur when the rock is under higher pressures and temperatures than those that exist at the surface. If I hand you a piece of limestone, and you try to bend the rock with all your strength, it may break, but it will not fold. Plastic deformation can only occur deep within the Earth where the temperatures and pressures are high enough for the rock to fold when stress is applied to it over long periods of time. What we see in a fold-thrust belt is both folding and faulting, however. How do these two processes occur together? I usually model this process to students by picking up a stick. When I start to apply a compressive stress to the stick, it will begin to bend (fold). Once I apply too much stress, however, the stick can no longer accommodate the strain by bending and it will snap (brittle deformation). This analogy's not perfect because sticks are more elastic than rocks and the stick will mostly snap back into a straight shape after it breaks. Rock, however, is not so elastic so it stays folded. Think of the syncline we just discussed. It folded until the stresses exceeded the rock's strength and then a thrust fault developed pushing some of the layers upward and out of the core of the syncline.

Before we move on to other features in the HVB, I'd like to point out that rocks in this outcrop along the Route 23B ramp are very fossiliferous. If you visit, be sure to take some time to examine them since they're full of brachiopods, honeycomb corals (*Favosites helderbergensis*), crinoid stems, and bryozoans. I've even found a nice marine snail (possibly *Holopea*) near here. Please don't collect, however, since this is a classic field trip stop for van-loads of geology students every semester and we would like them to have the pleasure of discovering fossils there as well.

These fossils represent organisms that lived in the sparkling waters of the Helderberg Sea. This sea was a perfect place for marine invertebrates with its shallow, sunlit, and clear waters. For millions of years, this area was a filter-feeder's paradise but things, as always, were about to change. The earliest signs that something was afoot were some clays that appeared in the formerly pristine water. The bottom sediments became muddier. It was more difficult for filter-feeders to exist as the turbidity increased.

What happened is land to the east was being pushed upwards again and forming areas of dry land. The land was subjected to weathering and erosion resulting in increased sediments entering the sea. Carbonate sediments on the seafloor became covered by clastic sediments. This is reflected in the geologic record by limestones grading upward into argillaceous (clay-rich) limestones and then into shales. This is exactly what you'll see if you continue driving west on Route 23; a sequence of limestones followed by shales a few kilometers up the road.

As we move from the Middle Devonian into the Late Devonian Period, we have another mountain-building event – the Acadian

Orogeny. The name is derived from an area of early French settlement in the Maritime Provinces of Canada. This is appropriate since the Acadian Orogeny did greatly affect this region but it's important to realize that it also affected virtually all of New England as well. As a matter of fact, much of easternmost Canada along with eastern Connecticut, eastern Massachusetts, Rhode Island, and coastal Maine were not even a part of North America prior to this mountain-building event. They represent a microcontinent that geologists call Avalonia that formed on the other side of the ancient Iapetus Ocean.

The microcontinent of Avalonia, named after the Avalon Peninsula in Newfoundland, had a complicated history before it accreted onto Laurentia to form much of what we now know as New England. Avalonia, probably composed of a number of different blocks of crust itself, was once attached to the great southern supercontinent of Gondwana in the latter part of the Proterozoic Eon (the time when Rodinia was rifting apart). It then broke free and drifted northward in the early Paleozoic Era before colliding with Baltica – a block of crust that's now Western Europe. Then Avalonia, along with Baltica on its opposite side, collided with Laurentia resulting in the Acadian Orogeny. In Western Europe, there was a similar mountain-building event but over there it's called the Caledonian Orogeny (from the Latin name for Scotland since it heavily affected the northern part of the British Isles).

While the Avalon terrane is often spoken of as a microcontinent which implies one large landmass, it's probably better described as an archipelago, similar to the chain of islands that make up the modern country of Japan. If you would like a modern analogue for the Acadian Orogeny, imagine what might happen if a subduction zone opened up in the Sea of Japan. As the oceanic crust subducted, Japan would move closer to the Asian mainland and eventually collide. This collision of the Japanese islands with the coastlines of Korea, China, and Russia would be an orogenic event similar to what happened when Avalonia collided with Laurentia during the Late Devonian Period. Japan would then cease to exist as a series of large, volcanic islands, they would instead become part of the Asian continent. In geologic parlance, we would say that the Japanese terrane docked with Asia in the Cenozoic Era. Similarly, we can say that the Avalon terrane docked with Laurentia in the Devonian Period.

The collision of Avalonia with Laurentia was not simple and occurred first in Canada and then progressively moved southward into the New England area over a span of time ranging from roughly 425 to 375 million years ago. The attachment of Baltica, Avalonia, and Laurentia formed a new continent called Euramerica which existed for over 100 million years.

As with the earlier Grenville and Taconic Orogenies, the Acadian Orogeny occurred in stages and had a long and complicated history. When tectonic plates are moving at only a few centimeters a year, it takes time to crush two continents together. This collision resulted in substantial mountains which, by the Late Devonian Period, rivaled the modern-day Rockies or European Alps in size. We'll discuss the erosion of these mountains in the next chapter.

The Acadian Orogeny left its mark to the east of the Hudson Valley in New England. Metamorphic rock and igneous intrusions have radiometric dates placing them in the Middle Devonian Period. Analysis of the mineral assemblages found within the metamorphic rocks indicate burial depths of over 20 km (12.5 miles) consistent with them forming deep in the core of large mountains. On the other side of these Acadian mountains were the sedimentary rocks of Avalonia. Cambrian Period trilobite fossils found in the Avalon Terrane of Massachusetts and Newfoundland are exactly the same as those found in Wales and England but entirely different from those found in the Hudson Valley. That's because during the Cambrian Period, the Avalonian crust underlying eastern Massachusetts was much closer to England than it was to the Hudson Valley far across the Iapetus Ocean.

It's worth the time to continue westward on Route 23 from the Route 23B entrance ramp where we left off our previous discussion. Beyond the syncline and small-scale thrust fault is a wooded gully without any rock outcrops. Then, as you start to pull off the ramp and onto Route 23, an imposing exposure of rock cliffs are seen. Stop your car on the wide shoulder, get out, and walk over to the outcrop. You should experience déjà vu. You will see Ordovician shales, the Taconic Unconformity, and the overlying Rondout Formation and Helderberg Group carbonates all over again. A complete repeat of strata that was seen a few hundred meters to the east. When I bring students here, we make these observations and then I ask them to try and figure out the geometric relationships that would make this possible. While it's challenging for introductory students to figure this out, it also becomes quite obvious when someone shows you the rather simple solution. It turns out that you're looking at the effects of folding. The wooded valley you just passed is the eroded core of an anticline, an upward-arching fold. The new exposure of the Taconic Unconformity exists because this contact is itself folded into a train of synclines and anticlines.

Walking west on the side of Route 23 will bring you past the Taconic Unconformity to the highly-deformed Rondout Formation. The narrow eroded gully on the outcrop here is another fault; faults are often areas where physical and chemical weathering can act more effectively since the rock is already broken. Note all of the small-scale folding next to the fault as rocks were dragged along this contact when the fault was actively moving.

The Taconic unconformity revisited. Ordovician Austin Glen sandstones and shales to the right and limestones of the Helderberg Group to the left.

Past the Rondout Formation, you will once again see the Helderberg Group limestones with the addition of the distinctive faintly-pink, coarsely-crystalline limestone of the Becraft Formation. Note how these limestones are practically vertical at the east end of the outcrop and flatten out to near horizontal at the west end near the bridge over the Thruway. Note also the large, flat surfaces, sometimes covered with layers of calcite and fault striations, at various locations along the outcrop. These are, of course, more thrust faults along where rock layers have slid during deformation in this area.

One common misconception in geology is the mistaken idea that faults are only found in seismically-active regions like California. It's not uncommon, when pointing out the numerous faults in the area, for students to ask why we don't have a lot of earthquakes here in the Hudson Valley. It's because these structures formed hundreds of millions of years ago during very different geologic settings. Those were times when the Hudson Valley was an active plate margin unlike the quiet passive margin existing today. During the Devonian Period, these faults may well have generated earthquakes when they slipped. Today, however, they're simply silent reminders of our sometimes violent past. That doesn't mean the Hudson Valley is immune from earthquakes, far from it as we've already discussed, simply that they aren't generated anymore by the ancient thrust faults we see in these roadside outcrops.

Large anticline (arching fold) cut through by a thrust fault (lower left to upper right) next to the Catskill Creek on Route 23. This is why this series of rocks is referred to as a fold and thrust belt.

Beyond the Thruway overpass, additional rock outcrops along Route 23 expose another sequence of heavily folded and faulted limestones of the Helderberg Group. This series of outcrops ends with a large anticline next to the bridge over Catskill Creek. On the western edge of the anticline, the near vertical beds of limestone dive down into the Catskill Creek far below. If you're physically fit (it's quite the climb back up), hike down to the creek where you can walk on the vertical beds of the Becraft Formation. Even a cursory glance at these rocks will reveal that they're jam-packed with brachiopod fossils. A former student even picked up some nice trilobites from here as well.

Another large thrust fault controls the orientation of the Catskill Creek through this area. It's not uncommon for streams and faults to coincide since, as mentioned previously, faults break and weaken the rock allowing it to become more easily eroded. The shale cliffs on the opposite side of the creek are the Esopus Shale; formations normally found between the Becraft Limestone and Esopus Shale are missing here due to this fault. The Esopus Shale was discussed in the previous chapter where it was found along Route 199 near Kingston. Driving westward on Route 23 past the Catskill Creek will reveal increasingly undeformed sedimentary rocks with essentially horizontal orientations. The HVB is a relatively narrow belt, only about 2.5 kilometers (1.5 miles) in width.

Rock layers going from horizontal (right) to vertical (left) along Route 23 in the Hudson Valley fold-thrust belt

These younger, horizontal sedimentary rocks west of the HVB are part of a sequence we see exposed throughout the Catskill Mountains. We'll discuss these rocks in the next chapter. A reasonable question to ask in reference to these rocks is why they are not deformed like those in the HVB. There are essentially two possibilities. The first possibility is that they're too young. These horizontal sedimentary rocks represent debris shed off the Acadian Mountains. If the HVB was deformed during the Acadian Orogeny, these rocks were deposited after the mountain-building essentially ended. The other possibility is that they simply formed too far from the mountain belt. In this case, the HVB could have been deformed by either the Acadian or the later Alleghenian Orogeny but deformation quickly died out as we move a few kilometers to the west of the Hudson River and these horizontal sediments were unaffected.

It's difficult to establish which of these two cases is correct because erosion has removed a lot of Devonian rock near the Hudson River. Becraft Mountain, on the east side of the Hudson River in Columbia

County, is an outlier of Helderberg Group rocks deformed just as they are in the HVB. These rocks are not physically connected to those west of Catskill because the intervening rocks have been eroded away. Similarly, the flat-lying Catskill Mountain sedimentary rocks once extended further to the east. Were they folded where they once occurred over the HVB? We don't know. If they were, that would tell us that the whole area was deformed by the Alleghenian Orogeny and not the Acadian.

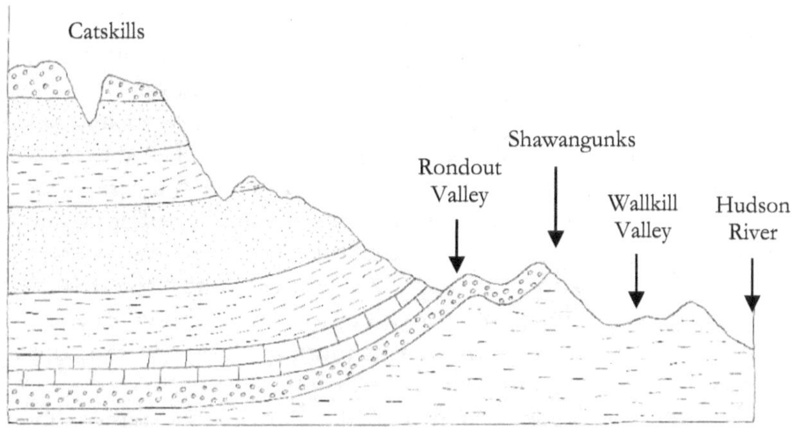

West to east cross section from the Catskills to the Hudson River. In a general sense, dotted patterns represent conglomerates, stipples are sandstones, dashed are shales, and brick patterns are limestones.

As mentioned earlier, I happen to believe that the HVB formed from the Acadian Orogeny. The deformed rocks are the right age and they are parallel to geologic structures of definite Acadian age not too far to the east in Vermont, Massachusetts, and Connecticut. The HVB is exposed in rocks from the Albany area southward to Kingston where things get a little complicated. Rock outcrops in Kingston show evidence of two different deformation events. One set of geologic structures parallels the HVB to the north. Another set of geologic structures, however, parallels structures mapped into the Rondout Valley which runs southwestward from Kingston toward Port Jervis (on the west side of the Shawangunk Ridge). In other words, structures north of Kingston were deformed by compression from almost due east and structures southwest of Kingston were deformed by compression from the southeast. This type of bend in a fold-thrust belt is called an orocline.

What does this change in orientations mean? Stephen Marshak, the geologist who mapped the HVB, and some of his colleagues have suggested that rocks in the Kingston area are recording the intersection of two separate episodes of deformation. The deformation of the HVB north of Kingston may well have occurred during the Acadian Orogeny while deformation southwest of Kingston in the Rondout Valley was likely to have occurred during the later Alleghenian Orogeny. Other structural geologists have argued that both the HVB and Rondout Valley structures were deformed by two separate phases of compression during the Alleghenian Orogeny and that the Acadian Orogeny resulted in no substantial deformation here in the Hudson Valley. At present, there's no way to settle this disagreement since folds and faults in sedimentary rock can't be easily assigned absolute ages.

Discussions in front of these outcrops by geologists often turn into what we call "arm waving" – a semi-derogatory term referring to those who make grand claims about the geologic evolution of an area based on very slim evidence. The expression comes from the fact that geologists often make sweeping gestures with their arms when talking about the collisions of ancient continents and the formation of long-vanished mountains. Part of a geological education is knowing how to distinguish between what we know to a high degree of certainty – the deformation of these rocks within a fold-thrust belt occurred during an episode of collisional tectonics – versus what is more speculative – the collisional event was the Acadian Orogeny.

Regardless of which orogeny folded and faulted the Siluro-Devonian carbonates in the Hudson Valley, we do know for a fact that this area was affected by the growth of the Acadian Mountains further to the east. Once the Acadian Mountains began to rise, they also began to erode away. Sediments from these eroding mountains were shed to the west and resulted in a thick wedge of sediment that is now seen exposed as the sedimentary rocks of the present-day Catskill Mountains – the subject of our next chapter.

9
Floodplains and Forests

Fossils have been long studied as great curiosities, collected with great pains, treasured with great care and at a great expense, and showed and admired with as much pleasure as a child's rattle or a hobby-horse is shown and admired by himself and his playfellows, because it is pretty; and this has been done by thousands who have never paid the least regard to that wonderful order and regularity with which Nature has disposed of these singular productions, and assigned to each class its particular stratum.

William "Strata" Smith
Notebook Entry

William Smith was an English canal surveyor who's often credited with being the father of stratigraphy – hence his nickname, "Strata Smith." Smith's contribution to the science of geology was simply to pay close attention to the rocks he was walking across as he surveyed routes for canals across England and Wales during the late 1700s. He learned to recognize many distinctive rock units – which today we would call formations – and created the first geologic map of England and Wales. This map was so significant that some have dubbed this accomplishment "the map that changed the world." This also happens to be the title of an excellent book on Smith by Simon Winchester if you'd like to learn more about his interesting, yet tragic, life (complete with an insane "nymphomaniac" wife and a stint in debtor's prison!).

William "Strata" Smith (1769-1839)

William Smith also noted that fossils throughout England and Wales occurred in very characteristic sequences which allowed them to be used in determining the relative ages of sedimentary rocks across widely separated areas. Smith was one of the first to view fossils not simply as attractive collectibles in the rocks, but as biostratigraphic tools. This observation, today called the Principle of Faunal Succession, is a obvious concept to anyone familiar with fossil collecting. Smith noted that certain fossil species always followed other fossil species in the stratigraphic record, never the other way around. This makes sense when you think about it – fossils represent organisms which live for a period of time and then become extinct. This happens over and over again for vast numbers of species over incomprehensibly long spans of time. The neat thing about this, however, is that for any particular period of time (the Middle Devonian Period, for example), there are certain characteristic assemblages of fossil species present within the rocks. If we find that same assemblage of fossil species in some other area, we can reasonably assume the rock formed during the Middle Devonian.

This observation, obviously, has important implications for the evolution of life on Earth. Keep in mind, however, that Smith was working over 50 years before Charles Darwin published his groundbreaking book On the Origin of Species in 1859. The Principle of Faunal Succession is also one of the strongest arguments against the young-Earth creationist idea of a global flood depositing all of the sedimentary rocks and fossils. If this were true, then all life preserved as fossils would be jumbled together in the stratigraphic record. Instead we find animals that lived in similar environments, brachiopods for example, sorted such that individual species are only preserved in very specific stratigraphic horizons everywhere in the world where they're found. *Gypidula coeymanensis,* for example, is only found in Helderberg Group age rocks. Never stratigraphically above them, never stratigraphically below them. Similarly this applies for thousands and thousands and thousands of other examples of fossils throughout the geologic record.

Even though we're two hundred years and thousands of miles removed from William Smith's England, we can see a distinctive faunal succession of fossils right here in the Hudson Valley. In Cambrian rocks, we see distinct assemblages of brachiopods, stromatolites, and trilobites. Ordovician rocks contain a number of different types of graptolites. In Silurian rocks, eurypterids are found. Devonian rocks contain a lot of brachiopods and tabulate corals. Even though certain groups of organisms, like brachiopods, are seen throughout the Paleozoic Era rocks, very different species of brachiopods are seen in Cambrian, Ordovician, Silurian, and Devonian Period rocks.

One of the exciting things about Devonian fossils, which are well-represented in New York State, is that they reveal several important advances in the history of life. The Devonian is sometimes termed the "Age of Fish" because vertebrate fish were common in Devonian seas. As a matter of fact, the Devonian Period had a higher diversity of fish than we do in today's oceans —there were five major classes of fish while today we only have three (and one is uncommon). Fish were also moving from the marine environment into freshwater settings. These freshwater fish, living in Devonian Period streams and swamps, are the same animals that evolved into the earliest amphibians.

The Devonian is when life became firmly established on land. While primitive plants and insects were appearing in a few isolated areas during the Silurian Period, full-blown forests existed in the Devonian and amphibians prowled these forests by the end of this Period. Some of these fossil forests are actually found west of the Hudson Valley in the middle of the Catskill Mountains.

Between Kingston and Albany, the Catskills dominate the western skyline from the Hudson River. While the Catskills are the most mountainous-looking feature in this part of the Hudson Valley, geologists actually don't consider them to be mountains at all. Unlike most mountain belts, the Catskills are composed of essentially flat-lying sedimentary rock. True mountains typically have highly deformed sedimentary rocks (as we saw in the Hudson Valley fold-thrust belt), metamorphosed rocks, and igneous intrusions – all of which are lacking in the Catskills. These so-called mountains are better described as a dissected plateau. They're essentially a big wedge of sedimentary rock which has been eroded into mountainous shapes by streams and glaciers carving deep, narrow valleys throughout the plateau.

Catskill peaks from the top of Overlook Mountain

The sedimentary rocks which form the Catskills are the erosional remnants of the Acadian Mountains. Streams and rivers drained off the western flanks of the Acadians and traveled westward across what's now New York State to an inland sea to the west. Throughout the Devonian Period, a new global marine transgression was occurring. We've discussed the Sauk (Cambrian) and Tippecanoe (Ordovician/Silurian) transgressions earlier; this new transgression is called the Kaskaskia and peaked during the subsequent Mississippian Period. While the Acadian Mountains rose from the sea in the East, much of the Midwest was submerging beneath the deepening Kaskaskia Sea.

Let's take a look at the stratigraphy of the Catskills. In Chapter 7, we left off with the Middle Devonian Onondaga Formation limestones which formed in a widespread sea covering New York (the limestone actually extends as far west as Michigan). In the upper part of the Onondaga are the Tioga ash beds which radiometrically date to about 400 million years ago. These ash falls represent volcanic activity recording the close approach of Avalonia with Laurentia. Avalonia was approaching because seafloor was being consumed down a subduction trench between the two continents. Subduction zones are always associated with volcanoes above the subducting plate.

With the collision of Avalonia, and subsequent growth of the Acadian Mountains, the area on the continental side of the mountains, known as a foreland basin, was apparently down-warped and the shallow carbonate sea which formed the Onondaga Limestones now became a deeper water basin with anoxic (oxygen-poor) bottom conditions. We see this reflected in the deposition of a unit called the Bakoven Shale on top of the Onondaga Limestone (further west in New York, the Bakoven grades into the better-known Marcellus Shale).

The Bakoven is a black, finely-laminated shale with few fossils. The big influx of clays to form the shale was from the erosion of the rising Acadian Mountains to the east. The black color of the shale is due to a high organic content. Shales which form in anoxic conditions build up high organic contents because the material is not broken down by bacteria in this type of environment. Bacteria require oxygen to break down organic detritus into its constituent chemicals. The low oxygen content in the bottom waters also means that scavengers and detritus feeders are generally not present. The few fossils in the Bakoven are generally planktonic organisms that, in life, floated up higher in the water column near the well-oxygenated surface waters.

There are a number of widespread, organic-rich, black shales that formed on the Laurentian continent during the Late Devonian Period. One particularly widespread unit in eastern North America is called the Chattanooga Shale which extends across numerous states just west of the present-day Appalachians. The character of these shales has been somewhat of a puzzle to geologists because they seemed to have formed in a type of environment we don't see in today's world. In the present day, black shales are forming in deep, restricted-circulation basins like the Black Sea north of Turkey. The Black Sea has a narrow connection to the Mediterranean through the Straits of Bosporus at Istanbul and any oxygenated marine or fresh water flowing into the Sea forms a "lid" above the deeper anoxic waters below. The sediments at the bottom of the Sea are black, organic-rich muds completely devoid of animal life. Pyrite is also a common mineral seen in these muds as well as in ancient black shales. Pyrite is iron sulfide (FeS_2) and is indicative of oxygen-poor environments; if oxygen is present, rusty red iron oxide called hematite (Fe_2O_3) will form.

The Devonian seas forming black shales were completely unlike the Black Sea, however, and show evidence of being open, relatively shallow seas, not deep, restricted-circulation basins. Geologists believe that the answer lies in the different environmental conditions that existed in Laurentia during the late Devonian. At the time, what's now eastern North America was only a few degrees south of the Equator. The area around the equator in today's oceans is sometimes referred to as the doldrums, a term from the days when sailing ships were becalmed there for weeks at a time because of the lack of winds. During the Late Devonian, the Kaskaskia transgression flooded large areas of Laurentia. The water was calm with little in the way of ocean current circulation and was receiving clay sediments from the Acadian Mountains to the east. The global climate during this time was warm; it's sometimes referred to as a greenhouse climate. There may have been no glaciers on Earth during this time, which would also account for the globally-higher sea level.

The significance of a greenhouse climate is that very cold ocean water doesn't form near the poles during these times. In our modern oceans, these colder and denser waters sink and drive deep circulation, called thermohaline currents, which help keep the deeper ocean waters well-oxygenated. In a greenhouse world, it's thought that the oceans would become more stratified and stagnant. This would result in bottom waters becoming anoxic over time. The Acadian Mountains to the east may also have acted as a barrier to oceanic circulation thus helping to maintain the stagnant waters.

While the bottom of these seas had virtually no life, it's likely that extensive blooms of plankton, and animals which fed on them, may have floated in the hot, sunlit surface waters of these continental seas. When they died and sank through the water column, such plankton would also provide the organic material for the bottom sediments along with the few fossils we do find in these shales today. Hartwick College geologist Bob Titus likes to refer to these epicontinental bodies of water as the "Poison Seas," an apt name for this hostile environment.

Following the black Bakoven Shale in the Hudson Valley are the brownish-red shales and sandstones of the Mount Marion Formation. Unlike the Bakoven, the Mount Marion sediments accumulated on a relatively-shallow, well-oxygenated seafloor and contain an abundance of fossils. For whatever reason, environmental conditions were changing and the ocean covering our area became less stratified. It may well have been due to the continued growth of the Acadian Mountains to the east and a change in sea levels.

The Mount Marion Formation, named after a town just north of Kingston, belongs to a larger group of rock formations called the Hamilton Group – thus the name Hamilton Sea applied to this shallow body of water covering New York. We are now starting the thick pile of sedimentary rocks which will eventually form the Catskill Mountains. While a little muddier and darker than the sparkling clear waters of the Helderberg Sea , the Hamilton Sea did support a wide variety of marine invertebrates including the ever-present brachiopods, a number of different trilobite species, horn corals, and mollusks such as clams and marine snails.

The Mount Marion Formation is very well exposed near Kingston and to the north. Route 209, north of Route 28, passes through a thick sequence of the Mount Marion. If you're in the southbound lane, there's a nice place to pull off and examine these rocks in detail. Note the thick shale units interrupted by thinner beds of sandstone. The small pits are from local residents collecting loads of shale for their driveways. It's hard to find fossils at this location, but there are areas along the cliff, within walking distance, where numerous horn corals and brachiopods weather out of certain layers and accumulate on the lower shale slopes.

Horn coral (Hyliophyllum halli) from the Mount Marion Formation

Route 32 near Saugerties also passes through a nice sequence of the Mount Marion Formation. Just north of the southbound Thruway entrance is a sequence of Onondaga Limestone with its distinctive "knobby" weathering. About 3 kilometers (2 miles) up the road, near Old Kings Highway, is a poignant historical marker for Indian Cave, the "home of Nachte Jan, last Indian of the region." I've been told that the cave in the Onondaga Limestone was destroyed by road construction. Route 32 then dips down and begins to climb a long hill. The outcrop on the side of the road is the overlying Mount Marion Formation. This area was once a town called Quarryville for the extensive bluestone quarrying operations nearby – a topic we'll discuss in more detail shortly.

One of the most spectacular fossils from the Mount Marion Formation was collected from the Saugerties area along the New York State Thruway. It's a large slab containing dozens of impressions of starfish (*Devonaster eucharis*) and clams which is currently on display in the lobby of the New York State Museum in Albany. If you visit, look carefully at the exhibit and you'll see that one of the starfish was even caught in the act of feeding on a clam. This fossil slab records a minor catastrophe as the starfish were quickly buried by sediment. The most likely explanation is simply a large tropical storm moving through this area, some 388 million years ago, with bottom currents washing sediments over the uncomprehending starfish entombing them for our later viewing pleasure.

Starfish (Devonaster eucharis) fossil from the Mount Marion Formation on display at the New York State Museum

In order to better understand the subsequent sedimentary rocks which comprise the present-day Catskills, we need to discuss the geologic concept of facies some more. Remember when discussing the Helderberg Group of formations, we saw that the different limestones formed in slightly different environments? Let's imagine a limestone forming in a shallow marine sea just below the wave base. If we trace the limestone in one direction, we might see its character change to a more sandy limestone and then to a shallow marine sandstone forming closer to the shoreline. Following the layer further might then bring us to terrestrial beach dune deposits. If we trace that original limestone in the opposite direction, however, we might see it change to deeper water oceanic shales. It's essentially the same layer forming at the same period of time. The lateral change in rock type we see is simply due to the lateral change in environments. Geologists would say that these different rock types belong to different sedimentary facies.

There are three ways that geologists commonly refer to facies in sedimentary rocks. There are metamorphic rock facies as well, but they refer to different temperature and pressure conditions of metamorphism. For sedimentary rocks, the first is lithofacies referring to the lithology or rock type. We can say the limestone facies changes into a sandstone facies as we approach the shoreline. We can also refer to a biofacies. This is a separation of sedimentary rocks based upon their fossil assemblages. We might say that the coral-brachiopod biofacies changes to a clam biofacies as we approach the shoreline. In contrast to these descriptive uses of the term facies, geologists may also apply the concept in a genetic sense, referring to the rock's genesis or origin in some ancient environment of deposition. A carbonate shelf facies might grade laterally into a shallow marine sand facies. While this last sense of the word involves some interpretation, it is very helpful to geologists since it tells us about the environment in which these rocks formed.

There's an important complication to this relatively simple picture. The Earth is a dynamic place and we haven't yet looked at the effect of time. Over time, environments change as continents drift, sea levels rise or fall, and climates change. As time passes, and environments change, the facies we see in sedimentary rocks will migrate laterally. This is an important concept in the Catskill sediments where we see a gradation over time from deeper marine sediments to shallower marine sediments to shoreline deposits and finally to terrestrial river and floodplain sediments as the Acadian Mountains eroded and sediments filled the adjoining Hamilton Sea. This happened first in the eastern part of New York and then occurred further and further toward the west. This westward progression of environments over time led to complex facies changes in the rocks in the Catskills which pose a challenge for stratigraphers in correlating sedimentary rock units – and a challenge for students to understand exactly what is going on here.

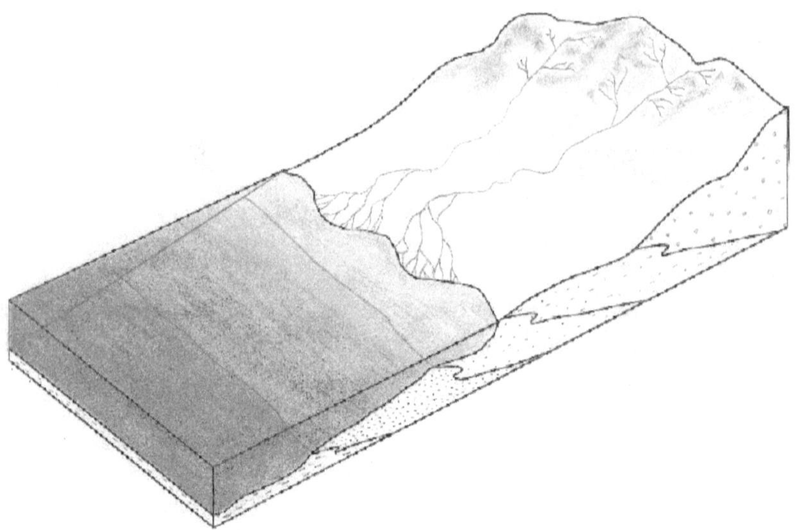

The concept of facies and facies change over time. Different sediments will be deposited in different environments (e.g. coarse sediments from mountain streams, sands on a beach, and clays in deeper marine settings). If sea levels rise or fall, or sediment deposition pushes the land out further into the sea, these environments, and the sediments they deposit, will migrate landward or seaward as well over time.

To simplify things, there are three basic Late Devonian sedimentary lithologies found in the Catskills – shales, sandstones, and conglomerates. Sedimentary structures within these rocks – like ripple marks, cross-bedding, mudcracks, and others – give us important clues as to the type of environment where these rocks formed. Fossils are even more important and often provide conclusive information as to the paleoenvironment. Finding brachiopods in a shale tells us it formed in a marine environment and not on a river floodplain, for example.

The lithologies, sedimentary structures, and fossils within the rocks of the Catskill Mountains tell us that they formed in several basic depositional environments. What we see is a progression from carbonate seas (Onondaga Limestone), to restricted circulation ocean water (Bakoven Shale), shallower marine sediments (Mount Marion Formation), shoreline deltaic deposits, meandering river and floodplain environments, and finally to braided streams at the foot of mountain slopes. The transitional (shoreline) and terrestrial deposits comprise what's known as the Catskill clastic wedge or Catskill delta – the 3.5 kilometer (11,500 feet) thick apron of sediments shed off the growing Acadian Mountains to the east.

We're not going to worry about the formational names for these rock units since even I have difficulty in telling some of these sandstones and shales apart from each other. What's essentially the same rock unit may have different names in the northeastern Catskills, the western Catskills, and the southern Catskills. What's more important is to learn to recognize the features in these shales and sandstones that give us important information about the ancient environments in which they formed. This ancient environment was very much like the modern environment of Bangladesh where the Ganges River and delta deposit vast amounts of coarse clastic sediments eroding from the high Himalayan Mountains.

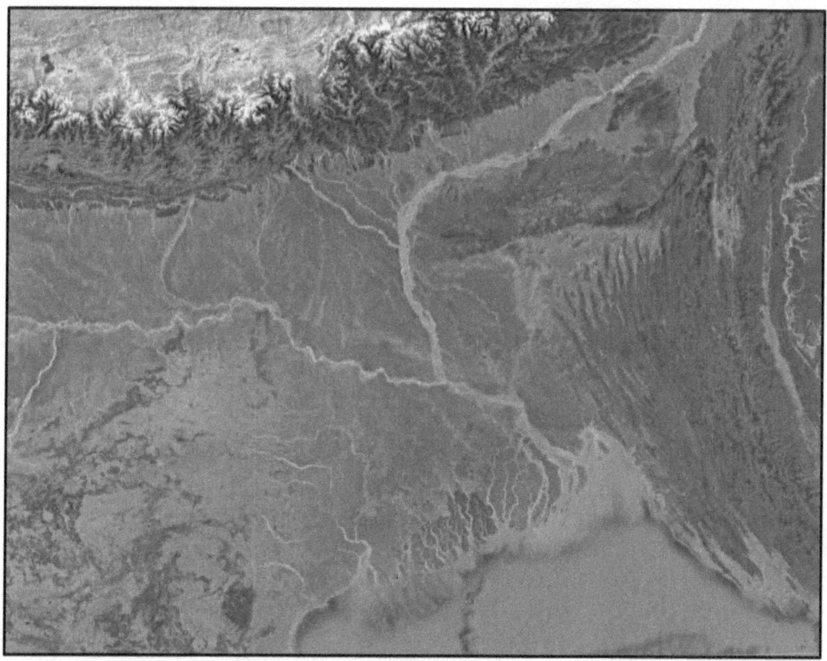

Rivers draining the Himalaya and forming the massive Ganges Delta into the Bay of Bengal. Analogous to the Devonian Catskill Delta draining the Acadian Mountains.

One of the most noticeable things in the Catskills is the red color of many of the sandstones and shales. These are called redbeds and are red because of small amounts of iron in the sediments. Since these were deposited in a terrestrial environment, there was plenty of atmospheric oxygen available to oxidize the iron into the rusty-red mineral hematite (Fe_2O_3).

The Devonian Period redbeds we find in the Catskills have something in common with Victorian England. In 1839, the year of William "Strata" Smith's death, British geologists Adam Sedgwick and Roderick Impey Murchison proposed the name "Devonian" for outcrops exposed in Devonshire. These Devonian rocks belonged to a famous and widespread rock formation called the Old Red Sandstone. While the Old Red Sandstone is not directly correlatable with the Catskill red sandstones, they did form at around the same time in Earth history and under very similar conditions.

The Acadian Mountains formed between Avalonia and Laurentia just south of the equator at the time. If we trace the Acadian Mountains northward across the paleo-equator, they essentially grade into the Caledonian Mountains, a chain which formed between Greenland and Baltica. Basically, the Old Red Sandstone was forming in Baltica, on the other side of the Caledonian Mountains from Greenland and Laurentia, much in the same way that the Catskill red beds formed. That area in Baltica is sometimes called the Old Red Sandstone Continent.

Laurentia during the Late Devonian (~360 Ma). The approximate locations of the equator and New York State are shown.

One excellent place to examine Catskill rocks is North-South Lake State Campground near Haines Falls in Greene County. The North-South Lake area is historically as well as geologically interesting. In 1824, the famous Catskill Mountain House was built on the nearby escarpment and had a spectacular view of the Hudson Valley. Three presidents and many of the leading political and business figures of the 19th century passed through this area. The abandoned and decaying Mountain House was burned in 1963 by the New York State Department of Environmental Conservation after they acquired the property.

The view from the former site of the Mountain House is one of the best in the Hudson Valley. While I normally deplore graffiti and the desecration of natural places, the sandstone ledges of the escarpment have a number of interesting old inscriptions carved into them by former guests of the hotel (please don't add to them). If you'd like to learn more about the old hotels in this area, I highly recommend The Catskill Mountain House by Roland Van Zant.

1902 photo of the Catskill Mountain House

If you're physically fit, take the time to hike along the Escarpment (blue) trail northward from the North Lake beach parking area (maps are available at the gatehouse). As the trail begins, take a close look at the sandstones upon which you're walking. Within the essentially flat-lying horizontal beds you will see other layers at an angle to the main beds. This is called cross-bedding and was mentioned previously when discussing the fluvial (stream-deposited) rocks of the Shawangunk Conglomerate. Cross-beds form from the migration of ripples in river bed sediments and are extremely common in Catskill sandstones. There are two basic types of cross-bedding – tabular cross-bedding where the cross beds are planar and trough cross-bedding where the cross beds are curved. Both types of cross-beds can form in river bed sediments and can be seen throughout the sandstones exposed in this area.

Cross-bedding in Catskill sandstone. Note the Swiss army knife for scale.

In addition to cross-bedding, some of the sandstones in the Catskills have thin, parallel, planar bedding. This planar bedding, typically formed in areas of fast-moving currents in the middle of larger rivers, allows the rock to be split into nice flat slabs unlike the cross-bedded sandstones. These occur throughout the Catskills and are called bluestone which is not really a geological term, and isn't very accurate since it refers to rocks that have a variety of colors but are mostly grayish. Bluestone mining used to be a big industry in the Catskills where many slabs were shipped to the Hudson River and then transported to New York City and elsewhere for curbstones and walkways. In the early 1800s, bluestone mining in the Catskills provided an impetus for the building of large numbers of Hudson River sloops to carry them to market.

While some bluestone is still mined locally in Ulster County, most of the quarries today are found in the southern Catskills of Delaware County. Since bluestone is typically not very extensive, it usually grades into cross-bedded sandstones laterally, the quarries are generally small operations where much of the work is still done by hand. It was always a business with a very small profit margin since mining often involved the removal of a lot of rock overburden to access the best layers. The availability of cheap cement for the building of sidewalks, steps, and curbs devastated the early bluestone industry.

If you'd like to see bluestone construction at its finest, visit Opus 40 between Saugerties and Woodstock in Ulster County. An eccentric artist named Harvey Fite spent 40 years of his life creating a massive outdoor environmental sculpture set in a bluestone quarry. A small Quarryman's Museum displays some of the tools used by traditional bluestone quarrymen. Two books on the Catskills and Woodstock by the late Alf Evers, an Ulster County historian, provide a fascinating glimpse into the life of those who locally mined bluestone for a living.

Opus 40 bluestone sculpture

Let's return to the Escarpment Trail. While climbing the small cliff near the beginning of this trail, note the thin layers of conglomerate within the sandstones. Conglomerate layers may represent pulses of uplift of the Acadian Mountains sending down coarser sediments into the adjacent fluvial environment. When the trail levels out, notice the joint sets on the flat, glacially-polished bedrock. In places, physical weathering processes like freeze-thaw cycles have widened some of the joints which now contain soil and vegetation. Note also that the cliff line is parallel to the joints; this is why the rocks which form the cliff broke where they did.

Sunset rock is a beautiful place to rest for a minute and enjoy the expansive view across the Hudson Valley. On a clear day, the towers of downtown Albany are seen to the north. As you look eastward toward the Berkshires, stretch out your arms and place one fist on top of the other. Imagine snowcapped mountains that high on the distant horizon. Those would have been the Acadians.

Continuing along the trail will bring you to a magnificent outcrop of conglomerate. This type of rock, which resembles traditional English Christmas or plum pudding, was called puddingstone in the Catskill Mountain House days. The rounded quartz pebbles in this cross-bedded conglomerate were clearly transported and deposited within a fast-flowing braided stream system. These types of conglomerates are common throughout the higher elevations in the Catskills and represent deposition in streams flowing off the eroding Acadian Mountains. Notice how the conglomerate is weathering away back into its constituent elements – sand and pebbles.

Catskill Puddingstone

A short distance away is the yellow spur trail which leads to Artist's Rock on top of the conglomerate outcrop seen along the blue trail. From this rock, you will have an excellent view back to North and South Lakes as well as the old Catskill Mountain House site. Before visiting this area, check out two paintings, both titled Catskill Mountain House, by Hudson River School artists Thomas Cole (1844) and Jasper Cropsey (1855). These paintings show this very same view 150 years ago. The human landscape changes, from generation to generation, but the rocks remain the same. We'll return to this area when discussing glaciation in the Hudson Valley.

Author at Artist's Rock with North-South Lakes in background. Painting is "A View of the Two Lakes and Mountain House, Catskill Mountains, Morning" by Thomas Cole (1844)

There is another interesting place to see Catskill-age conglomerates that's actually outside of the Catskills. Schunnemunk Mountain is a prominent ridge west of the New York State Thruway near Mountainville in Orange County. This ridge, sandwiched between the Precambrian rocks of the Hudson Highlands and Ordovician Martinsburg Shales of the Wallkill Valley, is composed of Late Devonian conglomerate. The Skunnemunk Conglomerate (pronounced the same, but spelled differently from the mountain), is a distinctive rock composed of rounded quartz pebbles in a purplish matrix. The conglomerate is almost a kilometer (3,000 feet) thick and, like the Catskill "puddingstones" to the north, represents braided streams flowing off the rising Acadian Mountains to the east.

People sometimes confuse the Skunnemunk Conglomerate with the Shawangunk Conglomerate about 40 kilometers (25 miles) to the west, but the Shawangunk Conglomerate represents fluvial (stream) deposits from the Taconic Orogeny around 420 million years ago and the Skunnemunk Conglomerate represents fluvial deposits from the Acadian Orogeny around 380 million years ago – that's a 40 million year difference and an entirely different mountain-building event. There are moderately strenuous trails to the top of Schunnemunk Mountain that are worth hiking to observe this interesting rock unit and for tremendous views of the Hudson Valley.

Schunnemunk Mountain is part of a complex, fault-bounded trough of Devonian-age sedimentary rocks that extends southward into New Jersey. The conglomerates seen here are likely erosional remnants of what once was a far more extensive deposit. Once you see the Skunnemunk Conglomerate, you'll start to recognize boulders and cobbles of it lying around throughout southeastern New York. These were glacially transported and one easily-viewed example is a large boulder of purple conglomerate next to the observation tower parking lot at the top of Bear Mountain in the nearby Hudson Highlands.

One enigmatic feature of the Catskills is known as the Panther Mountain structure. Geologists, examining maps of the Catskills, have long noticed a curious circular structure centered on Panther Mountain, a peak just east of Route 28 near Phoenicia. The circular feature is outlined by the path of two streams – the Esopus and Woodland Creeks. This is significant since stream patterns typically reflect something about the underlying geology of an area as they will preferentially erode paths where the bedrock is weakest.

There are basically three processes that can result in a circular feature – volcanic activity, a domical uplift, or a meteorite impact. Volcanic activity can quickly be ruled out since the Catskills are a pile of sedimentary rocks and there are no volcanoes or volcanic rocks anywhere in the area. Domical uplifts typically occur with the intrusion of igneous rock. Once again, there's no evidence of this occurring in the Catskills.

What about a meteorite impact? The problem with this idea is that Panther Mountain is obviously a mountain, not a meteorite crater. How can a meteor impact result in a mountain? The explanation is that the meteorite would have impacted this area before the Catskills even existed. Let's travel back in time to the middle part of the Devonian Period. Prior to the deposition of the sediments which formed the Catskills, there was a shallow sea in this area – the end stages of the Helderberg Sea mentioned earlier. A meteorite came out of space one fine morning and splashed into the sea excavating a crater on the seafloor. How big was this resultant crater? Using Google Earth, I measured the circular feature as being about 10 km (6.2 miles) in diameter.

How much energy would be released by such an impact? H. Jay Melosh, a geophysicist at Caltech, has published equations to calculate such things given a variety of factors (density of the meteorite, velocity of the meteorite, angle of impact, type of material impacted, etc.). Let's assume a nickel-iron meteorite with an average density of 7,800 kilograms per cubic meter, a conservative impact velocity of 20,000 meters per second (almost 45,000 miles per hour), and a vertical impact coming into the shallow Devonian sea and striking a limestone seafloor. Given these assumptions, the impactor would have a diameter around 250 meters (about 820 feet). A meteorite this size would impact with an energy equivalent to 1,800 megatons of TNT – the combined energy of over 138,000 Hiroshima-sized (13 kiloton) atomic bombs!

This obviously would have been a traumatic event for the marine invertebrates living in that sea and would have likely affected the global climate for a time. Such impacts shoot up a lot of debris into the stratosphere where it's spread by jet-stream winds to eventually encircle the Earth. This dust will then reflect some of the Sun's energy away resulting in a several-year-long cooling of the Earth's atmosphere. This is sometimes called the "nuclear winter" effect after a number of studies on the effects of a (hopefully) hypothetical large-scale nuclear war; an event which would also throw significant amounts of dust into the stratosphere. Over the past few decades, researchers have recognized more and more ancient meteorite impacts in the geological record leading us to wonder when, not if, we'll be hit again.

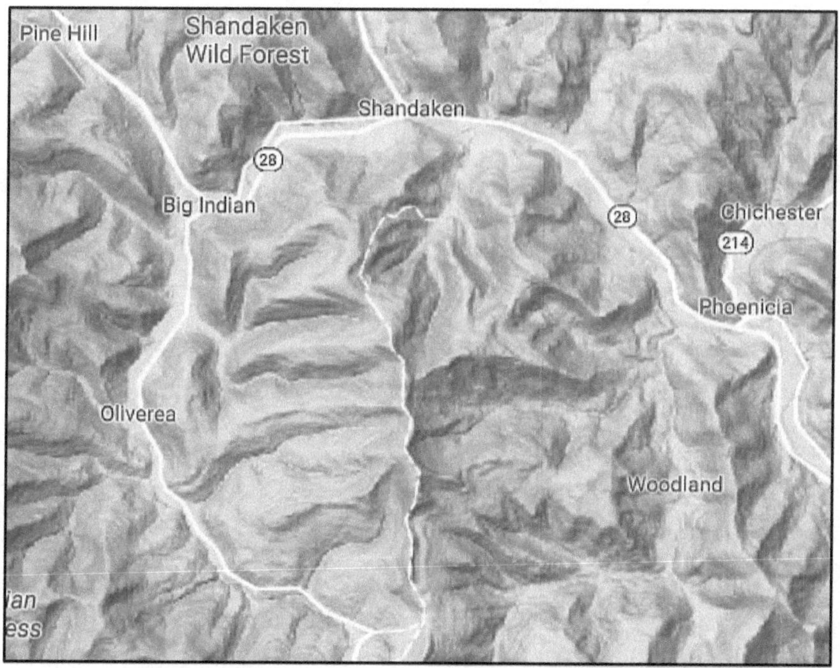

Circular feature around Panther Mountain (center) near Phoenicia

This meteorite impact would have resulted in a circular system of fractures in the bedrock outlining the crater. As the Acadian Mountains grew, and then eroded away, this meteorite crater eventually became buried under kilometers of sediments – the sediments which eventually became the rock layers of Panther Mountain. With more recent erosion, the Woodland and Esopus Creeks preferentially weathered away bedrock along joints outlining the position of the ancient crater about 350 meters (1,150 feet) below the streambed.

How much hard evidence is there for this structure originating as an impact crater? While there's no definitive evidence, a number of highly suggestive clues have been presented by one of the biggest proponents of this feature, the late Yngvar Isachsen who was a geologist with the New York State Geological Survey. Isachsen hiked the course of the Esopus Creek and noted that joints in the bedrock here were ten times more abundant than in the surrounding area. Panther Mountain is composed of relatively flat-lying, undeformed sedimentary rock. The high fracture density calls out for an explanation and the presence of a buried crater rim could explain this feature. Isachsen also hiked over Panther Mountain with a gravitometer – a sensitive instrument which measures minute variations in the Earth's gravity field caused by density differences in the subsurface rocks. There is less of a gravitational attraction directly beneath Panther Mountain which implies lower density rock. One possible explanation is that lower density is the result of heavy fracturing of the subsurface bedrock within the meteorite crater.

The most compelling evidence, however, came in the mid-1990s from an examination of sediment samples taken from a natural gas well drilled in this area back in 1955. The samples, stored and forgotten in the New York State Museum, were rediscovered and tediously examined, grain by grain, under a microscope to eventually reveal tiny spheres of glass. The significance of this is that such spheres commonly form near meteorite craters. The tremendous kinetic energy released by impacts will melt rock and the molten droplets quickly solidify into a silica glass as they fly through the air and settle nearby. In addition, grains of shocked quartz were also found. These form from shock waves passing through the crystalline lattice of the mineral leaving behind parallel lines of melted and unmelted quartz. Both of these pieces of evidence strongly support the idea of an ancient meteorite impact here.

Another former New York State Geological Survey geologist, Donald Fisher, has suggested that a rock outcrop located along Route 54 east of Cherry Valley in Otsego County may have recorded this impact event. This outcrop, about 88 kilometers (55 miles) northwest of the Panther Mountain structure as the crow flies, contains an enigmatic layer of heavily deformed rock within the Union Springs shale. This deformation may have resulted from shock waves from the meteorite impact and, if this interpretation is correct, would date the impact at around 390 million years ago near the end of the Middle Devonian Period.

We can't leave our discussion of the Catskills without talking about its famous fossil forests. In the middle of the Catskills is a small town called Gilboa located along the Schoharie Creek. This was a good place for a town in the 18th century due to the creek flowing through a natural gorge allowing early settlers to use water to power their mills. Fossilized tree stumps were discovered in the Gilboa area in the early 1850s by an ordained minister and part-time naturalist named Samuel Lockwood and later named *Caulopteris lockwoodi* in his honor. In October of 1869, a large flood washed down the Schoharie Creek destroying many of the bridges. During quarrying of stone for rebuilding the damaged structures, more fossil tree stumps were discovered. This time, however, they came to the attention of James Hall, the State Paleontologist, and he sent out people to collect and bring back specimens for the New York State Museum in Albany.

The real bonanza came with the construction of the Schoharie Reservoir here around 1920. The reservoir is part of the system which supplies water to New York City and a number of additional fossil stumps were discovered during excavations for the dam. These fossils were intensively studied by a remarkable woman paleontologist, Winifred Goldring. Goldring held a variety of positions at the State Museum in Albany, eventually becoming State Paleontologist. She also became President of the Paleontological Society and Vice President of the Geological Society of America, all unusual honors for a woman in the early 20th century and attesting to her brilliance. In addition to the Gilboa fossils, Goldring studied the Cambrian stromatolites west of Saratoga Springs, produced a monumental work on the Devonian crinoids of New York, and wrote a guide to the geology of John Boyd Thacher State Park, among many other things. Goldring's research on the Gilboa fossils resulted in a famous diorama showing the reconstructed Gilboa forest during the Devonian Period. This diorama was a centerpiece of the New York State Museum for 50 years before it was retired.

More recently, construction of a hydroelectric facility north of the Schoharie Reservoir in 1969 resulted in the discovery of many more fossils. A lens-shaped rock unit within the Middle Devonian Panther Mountain Formation was discovered and it represented a freshwater shoreline deposit that was packed with plant material. The fine-grained mudstone showed exquisite preservation of leaves, stems, and spores from a type of plant (*Leclercqia complexa*) known as a lycopod. Lycopods are small and inconspicuous plants today. One group, the club mosses, can still be found on the shady floor of Catskill forests where they're sometimes called ground pines. Back in the Devonian Period, however, lycopods in the Catskill delta were tree-sized and formed forests along the many streams and wetlands of the area. Lycopods remained important during the subsequent geologic period, the Carboniferous as we'll see in the next chapter.

Outdoor display of fossil tree trunks near Gilboa

Diorama of the Gilboa forest shown on the sign above

In addition to a multitude of plant fossils, additional microscopic study of the material has led to the discovery of remarkable insect remains. There are a number of predatory arachnids (spider-like insects), centipedes, and even the fossilized remains of a spider spinneret. Since flying insects had not yet appeared in the fossil record, the silk was probably used to wrap egg sacs rather than to spin webs.

New discoveries of root systems in a quarry near the town of Cairo have pushed back the age of this forest to 386 million years ago.

The 380 million-year-old fossils from the Gilboa area reveal a complete ecosystem in the latter part of the Middle Devonian Period along the flanks of the Acadian Mountains. It was apparently an ecosystem of carnivores since herbivorous insects had yet to evolve. The stage was now set for the appearance of amphibians first arising a few million years later in the Late Devonian (amphibians like to eat insects – a good reason to come out onto land).

By 375 million years ago, organisms appeared which shared both fish and amphibian features and have been dubbed "fishapods" by some paleontologists. An example is a fossil called *Tiktaalik* recovered from Ellesmere Island in far northern Canada in 2004. *Tiktaalik* looked fish-like yet had fins with well-developed wrist bones and simple fingers which allowed it to support its own weight. Unlike fish, it also had a flat head and mobile neck and its ribs indicate that it had lungs. This fossil was discovered in stream-deposited sediments very similar in environment and age to those which formed the Catskills.

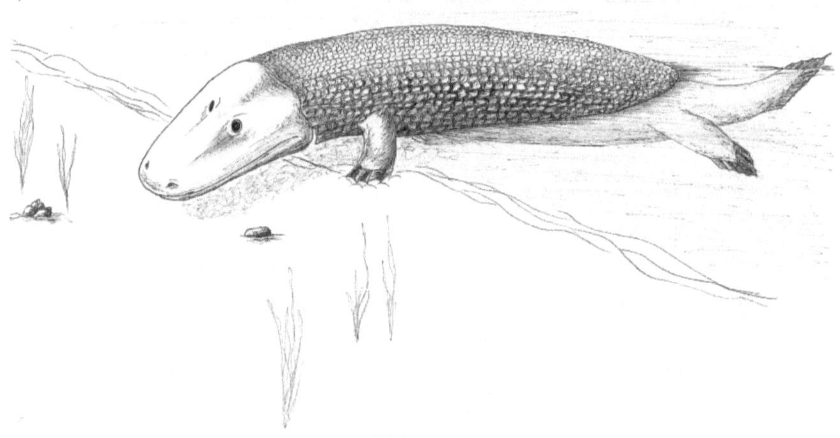

Tiktaalik

While true amphibians did not yet exist, lobe-finned, air-breathing fish were around during this time and probably appeared occasionally on the Gilboa mudflats even though we have no direct fossil evidence. The occurrence of fossil tree stumps in several stratigraphic levels in the area indicates that infrequent severe rainstorms would occasionally blast over the area causing flash floods. These would bury the bases of the trees in sediment and eventually kill them. One such storm apparently washed a tree out to sea where it eventually became waterlogged and sank down into the clay mud on the seafloor of western New York near the present-day town of Naples in Ontario County (south of Canandaigua Lake). In 1882, this 3.6 meter (12 foot) "Naples Tree" (*Lepidosigillaria*) was discovered and now resides in the collections of the New York State Museum in Albany.

The Acadian Mountains continued to weather away as streams and rivers relentlessly carried sediments through ancient forests to the sea. This long period of calm, however, was about to be interrupted by yet another major mountain-building event – the Alleghenian Orogeny. This is the collision which built the supercontinent of Pangaea, the subject of our next chapter.

10
Pangaea

The Appalachian Chain is the most elegant on earth, so regularly arranged that its belts of formations and structures persist virtually from one end to the other – from its first appearance from beneath the sea in Newfoundland, to its final disappearance under the Gulf Coastal Plain in Alabama... No wonder is it that the Appalachians have been the birthplace of many of the great principles of North American geology and of World geology... But the apparent simplicity of the Appalachian Chain is deceiving; actually, it is full of guile, and its geology has aroused controversies as acrimonious as any of those in our science.

Philip B. King
Studies of Appalachian Geology

Eastern North America has been affected by four major mountain-building events. The first, the Grenville Orogeny occurring between 1.3 – 1.1 billion years ago, created the very continental crust we're sitting on and resulted from the collision of Laurentia and Amazonia during the formation of the supercontinent of Rodinia. Next was the Late Ordovician Taconic Orogeny resulting from a volcanic island arc colliding with Laurentia. The third was the Acadian Orogeny, a collision of the Avalon Terrane with Laurentia during the Mid- to Late Devonian Period. The last mountain building event

187

occurred at the end of the Pennsylvanian Period and into the Early Permian Period as Africa collided with Laurentia resulting in the creation of yet another supercontinent called Pangaea. This last event is called the Alleghenian Orogeny, although you'll sometimes see it referred to as the Appalachian Orogeny.

The three orogenic events which occurred during the Paleozoic Era (Taconic, Acadian, and Alleghenian) formed what geologists today call the Appalachian Mountains. While most people think of Appalachia as an economically-depressed coal-mining region centered in West Virginia, geologists define the Appalachians as an orogenic, or mountain, belt stretching for over 3,000 kilometers (about 2,000 miles) from Alabama and Georgia up the East Coast through New England and then into the Canadian Maritime Provinces. While different names are applied to different parts of the Appalachians – Great Smoky Mountains, Blue Ridge Mountains, Allegheny Mountains, Berkshires, Green Mountains, White Mountains – they are really all part of the same belt of deformed rocks (albeit with differences in their geologic composition and histories). At various times in the past, these mountains were characterized by high, jagged peaks. Their lower, rounded shapes today attest to a long history of erosion and old age.

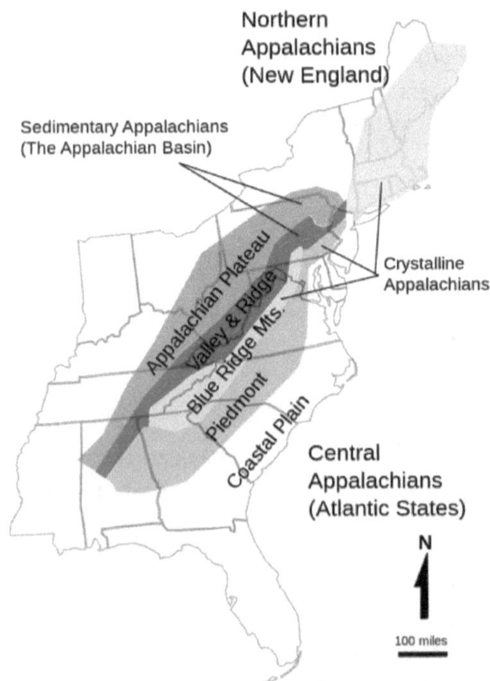

Geology of the Appalachians in Eastern North America

Philip B. King, a field geologist with the United States Geological Survey for many years, was correct in his assessment of the Appalachians in this chapter's epigraph. Despite an apparent simplicity, and its lack of "sexiness" to many geologists who would rather travel to more exotic locations, there are still many things to learn from studying the geology of the Appalachians. Even here in the Hudson Valley, where geologists have been working for over two centuries, we still have numerous unanswered geologic questions and opportunities for fruitful research for years to come.

When we left off in the previous chapter, the Acadian Mountains were eroding away to provide the thousands of kilometers of sediment for the Catskill clastic wedge in the Late Devonian Period. Life was now thriving on land, especially along the banks of rivers and coastal swamps in this tropical climate. Then, here in New York State at least, the recording ends. We have a gap of over 100 million years from which we have no rocks preserved. The only exceptions are a few scattered outcrops along the Pennsylvania border in western New York. We do, however, have an excellent record of Late Paleozoic rocks in Pennsylvania. What happened in our area?

New York was obviously around during this time in the Late Paleozoic Era and was primarily a terrestrial environment above sea level. Since we have rocks of this age exposed in Pennsylvania to the south and Ohio to the west, a number of geologists believe that rocks from the Mississippian, Pennsylvanian, and perhaps Permian Periods were also once present in New York State but have since been stripped off by erosion. There is some evidence to support this idea. When sedimentary rocks are buried, the increased temperatures and pressures, while not high enough to metamorphose the rock, will physically and chemically alter the mineral grains, any organic components, and various chemical isotopes within the constituent minerals. By examining these features, some geologists have argued that the Devonian rocks in the Hudson Valley region may have been buried several kilometers below the surface at one time. If this interpretation is correct, what were they buried under? The only reasonable answer would be younger sedimentary rocks that are currently missing from our area.

During the Alleghenian Orogeny, our area was once again uplifted and this would have accelerated rates of erosion removing all of the overlying Mississippian and Pennsylvanian Period sedimentary rocks. Sediments from these eroded rocks formed the present-day sedimentary strata of the Allegheny Plateau in western Pennsylvania and eastern Ohio. Sedimentary rocks which once existed in New York have been recycled into sedimentary rocks now exposed in Pennsylvania — all part of the great rock cycle that has been occurring on Earth for billions of years.

The Alleghenian Orogeny was the largest mountain-building event to affect our area since the Grenville Orogeny a billion years earlier. The Alleghenian Mountains may have been truly Himalayan in scale with peaks of 8,000 meters (26,000 feet) or more in elevation. The Taconic and Acadian Mountains, by contrast, were still significant but more analogous to the Rockies or European Alps in size with peaks probably below 5,000 meters (16,000 feet). It's difficult, however, to accurately estimate the size of mountains that have since weathered away. Reconstructions are based primarily upon two things. The first is the degree of metamorphism of rocks formed within the core of the mountain belt since these temperature and pressure conditions roughly correlate to depth of burial. Secondly, the height of ancient mountains can also be estimated by the total amount of sediments eroded from them and deposited elsewhere.

Approximate shape of Pangaea (~240 Ma) in the early Mesozoic Era

After the sandwiching of the Avalon Terrane between Laurentia and Baltica (northern Europe) during the Acadian Orogeny, the Iapetus

Ocean finally closed as Africa approached and collided with Laurentia during the Late Pennsylvanian Period. This, and other collisions with Asia to the north and collisions of South America, Antarctica, and Australia to the south, formed the supercontinent of Pangaea (also commonly spelled Pangea).

Pangaea is from the Greek for "all Earth" and this supercontinent was surrounded by a global ocean called Panthalassa (Greek for "all sea"). Pangaea was shaped somewhat like the letter "C" with the Tethys Ocean on its eastern side. The name Pangaea was first proposed by Alfred Wegener, the German climatologist who is generally credited with developing the concept of continental drift (the precursor to plate tectonics) in his 1915 book The Origin of Continents and Oceans as discussed earlier in Chapter 2.

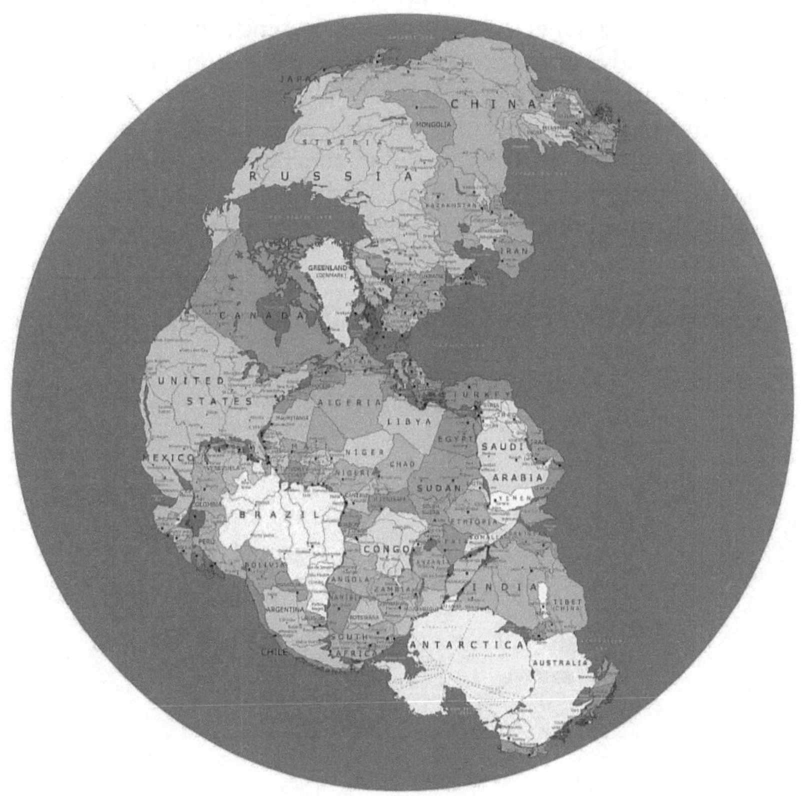

Pangea Politica — The supercontinent of Pangaea with modern-day countries superimposed.

Thus far, we've spent a lot of time discussing the ancient positions of the continents. What we haven't really done, however, is explain how geologists know where certain blocks of continental crust were at various times in Earth history. How do we know that Amazonia collided with Laurentia during the Grenville Orogeny? How do we know that Laurentia was in the Southern Hemisphere for much of the Paleozoic Era. How do we know Pangaea was a C-shaped supercontinent that formed at end of the Paleozoic? To many people, unfamiliar with geology, it seems like we're just making it all up.

It turns out there is abundant evidence to support these plate tectonic reconstructions. Let's look at Pangaea as an example. When Alfred Wegener first proposed the idea of an ancient supercontinent back in the early 20th century, there were several observations supporting this idea. The first was simply the apparent fit of the continents. Anyone looking at a world map today can see that Africa and South America look as if they had once been connected (and the fit is better when comparing the edges of their continental shelves – not the present-day shorelines – since they're the true edge of the continents). Not only that, but there are rocks and geologic structures in western Africa that exactly match in lithology and age with identical ones in Brazil. Evidence of ancient glaciations, in the form of glacial deposits and the directions of glacial striations (scratches in bedrock recording glacial movement), are best explained by proposing that the continents were once connected.

Glossopteris leaf fossils

Some of the most-compelling evidence for ancient plate connections, however, is in the form of fossils. One would expect to find different species of plants and animals in Africa than in South America since they're separated by some 5,000 kilometers (3,000 miles) of intervening ocean. Two famous fossils (at least to geologists) are the leaf

Glossopteris and the fresh-water crocodile Mesosaurus. Both represent organisms thought to be unable to cross ocean basins yet fossils of them are found in both South America and Africa (as well as other Southern Hemisphere continents like Antarctica and Australia similarly thought to be connected when Pangaea existed).

Mesosaurus fossil

While matching rock types and fossils can tell us certain blocks of continental crust were connected, they don't really tell us where these continental blocks were on the surface of the Earth. We may be able to recognize that certain rocks or fossils formed in specific paleoenvironments, and we can then make reasonable guesses about the latitudes at which we'd find those environments, but that still leaves a lot of leeway. Several decades after Wegener's passing, however, a new subdiscipline of geology known as paleomagnetism was developed. To understand paleomagnetism, we have to first discuss the Earth's magnetic field.

Way back in Chapter 1, it was mentioned that the Earth has a magnetic field due to the movement of liquid iron within the outer core. We can visualize this as a bar magnet inside the Earth with a north and south magnetic pole. That's why a compass points north, it lines up with the invisible lines of magnetic force connecting these two poles. One complication, however, is that the Earth's magnetic north and south poles don't exactly line up with the geographic north and south poles (axis of the Earth's rotation). That slight difference is the magnetic declination or variation of a compass's magnetic north from true north.

What most people also don't realize that if you had a compass needle that could freely swing in the vertical direction (up and down), it would show you the magnetic latitude. On the magnetic equator, half way between the magnetic poles (slightly offset from the geographic equator and poles), a compass needle would be horizontal. In other words, it would have an inclination of 0° when it's located at 0° of magnetic latitude. At the magnetic poles, which are at 90° of magnetic latitude, a compass needle would be oriented straight up and down, a 90° inclination, since it's aligning with the magnetic lines of force where they plunge directly into the Earth. Where I teach at SUNY Ulster, such a compass needle would have an inclination of 51° (our geomagnetic latitude, which differs a bit from our geographic latitude of 42°).

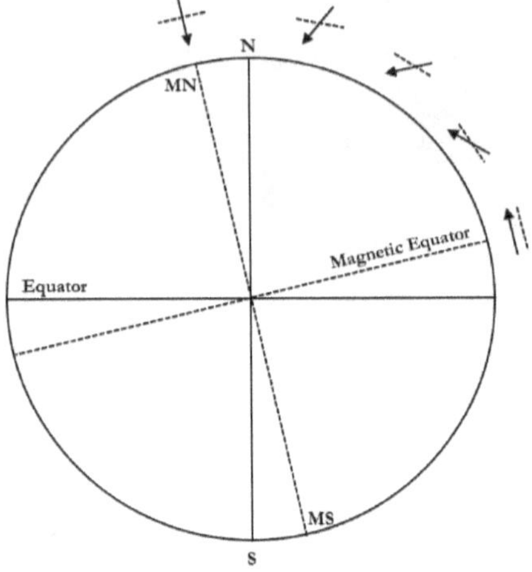

How the angle of inclination of a compass needle (arrows) changes with magnetic latitude. N and S are the North and South geographic poles and MN and MS are the north and south magnetic poles.

Why is this significant? Because some rocks can actually preserve the direction and inclination of the Earth's ancient magnetic field This fact is the basis of the science of paleomagnetism. Imagine a basalt lava flow. As lava cools, small magnetic grains of the aptly-named mineral magnetite (Fe_3O_4) will align with the magnetic field at the time of formation. Sedimentary rocks can also preserve evidence of the paleomagnetic fields since magnetic grains can align with it as they settle into seafloor sediments. If a block of continental crust containing that lava flow or sedimentary rock was in the Southern Hemisphere and rotated by 90°, as Laurentia was in the Paleozoic Era for example, then the magnetic field recorded in that rock will not align with the present-day magnetic field. The little mineral grain compass needles will be pointing in a different direction (reflecting the rotation of the continent) and have a different inclination (reflecting that the continent was at a different latitude when the rock formed). This allows us to reconstruct the ancient position of that continent.

In reality, there are a many additional complications to this simple picture. The Earth's magnetic poles wander slightly over time and the magnetic field even periodically reverses polarity. The paleomagnetism of rocks can be reset by later heating events (metamorphism) or groundwater flowing through the rocks and interacting with the magnetic minerals. Paleomagnetic data also only gives us latitudinal data, not longitude. If we discover through paleomagnetic studies that two continents were both at 30° south latitude at the same time, it doesn't tell us exactly how they may have fit together. With careful field work, comparison of many, many paleomagnetic samples of different ages from all over the world, and the assistance of other types of geologic data, however, geologists have built up what we believe to be an increasingly accurate reconstruction of the ancient world at various periods of geologic time. So when a geologist tells you that Laurentia was deep in the Southern Hemisphere in the Early Paleozoic Era, there's a lot of generally unseen evidence supporting that bold statement.

Let's now return to the time when Pangaea had formed at the end of the Paleozoic Era. The collision of Africa with Laurentia heavily folded and faulted sedimentary rocks that now form what geologists refer to as the Valley and Ridge Province. The Valley and Ridge is the fold-thrust belt from the Alleghenian Orogeny. The folds, trains of anticlines and synclines, thrust faults, and differential erosion of the sedimentary rock layers gives the Valley and Ridge its distinctive topography. If you've ever driven south on the ups and downs of Interstate 81 through Pennsylvania, you've experienced the Valley and Ridge firsthand.

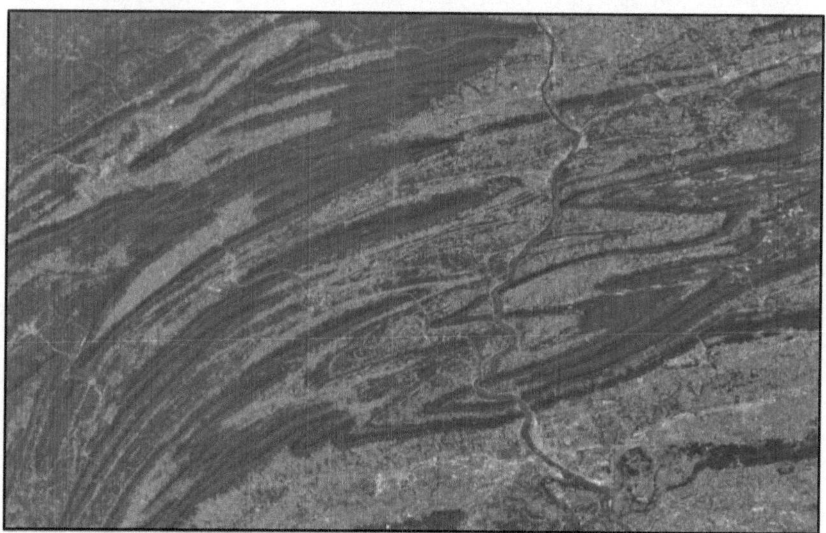

Satellite image of the Valley and Ridge (and Susquehanna River) in central Pennsylvania clearly showing a landscape formed by folded rocks.

195

The Valley and Ridge extends from southeastern New York into the northwestern part of New Jersey, through Pennsylvania, Maryland, Virginia, West Virginia, Tennessee, and Alabama. Earlier, in Chapter 8, we were introduced to the Hudson Valley Fold-Thrust Belt (HVB) and discussed the controversy regarding its time of formation. If the HVB formed during the Acadian Orogeny, it's not part of the Valley and Ridge Province. If the HVB formed during the Alleghenian Orogeny, it can be considered the northern end of the Valley and Ridge.

In southeastern New York, the Valley and Ridge is reflected in folded and faulted sedimentary rocks exposed in the Rondout Valley and the Shawangunk Ridge on its east. While the Shawangunk Conglomerate was deposited during the Silurian Period, and the limestones of the Rondout Valley formed in the Devonian Helderberg Sea, they were deformed during the Alleghenian Orogeny by the horizontal compressive forces from the collision of Laurentia with Africa.

We discussed the formation of the Shawangunk Conglomerate in Chapter 6. The Taconic Orogeny pushed up the Precambrian rocks of the Hudson Highlands towards the end of the Silurian Period. Erosion of these metamorphic rocks produced the pebbles and sand that washed westward toward the sea covering central New York at the time. They were, of course, deposited in horizontal beds in a fluvial environment. Today, however, the sedimentary beds of Shawangunk Conglomerate are folded, jointed, and cut through by both large and small faults. This deformation was a direct result of the Alleghenian Orogeny and the present-day topography of the Shawangunk Ridge is a result of later erosion, primarily glacial, sculpting these folded rocks.

To the west of the Valley and Ridge is the Appalachian Plateau. This is also referred to as the Allegheny Plateau in the north and the Cumberland Plateau in the south. It's all relatively horizontal sedimentary rock composed of debris shed off the eroding Appalachian Mountains and analogous to the sediments comprising the earlier Catskill Delta from the eroding Acadian Mountains almost 100 million years earlier (and the Queenston Delta shed from the eroding Taconic Mountains about 150 million years earlier).

The eastern end of the Valley and Ridge is called the Great Valley, a major geographic feature of eastern North America. Like the Appalachians, the Great Valley has different names along its 1,800 kilometer (1,100 mile) length from Alabama up to the Canadian Border. Notable areas are the Tennessee Valley, Shenandoah Valley, Lebanon and Lehigh Valleys, our own Hudson Valley, and the Champlain Valley. The Great Valley had been an important north-south route along the Appalachians from the earliest Native Americans to present-day

Interstates. Patterns of settlement in the early history of our country and the great battles of the Civil War were all heavily dependent upon the geography of the Great Valley and surrounding ridges and passes.

From Pennsylvania southward, the eastern side of the Great Valley is characterized by the Blue Ridge and Great Smoky Mountains, igneous and metamorphic rock primarily of Grenville age that were uplifted and exposed by the Alleghenian Orogeny. These mountains are similar in many respects to the Hudson Highlands in New York State – a topic we'll come to shortly. To the east of the Blue Ridge and Smoky Mountains is the Piedmont (French for "foot of the mountain"), a plateau characterized by low, rolling hills and underlain by a complex assemblage of heavily-eroded igneous and metamorphic rocks which formed during the Taconic, Acadian, and Alleghenian Orogenies.

To the east of the Piedmont is the Atlantic Coastal Plain, a flat-lying area of Cenozoic Era sediments deposited after the breakup of Pangaea. The boundary between the Piedmont and the Coastal Plain is called the Fall Line. The Fall Line is characterized by small waterfalls where rivers flowing off the more resistant Piedmont enter the lower-elevation coastal plain area. Many large cities, like Richmond, Virginia on the James River, developed as ports on the Fall Line due to difficulties in shipping goods further upstream at those locations.

Physiographic provinces of New York

While these geologic provinces are prominent in states like Virginia, they are also expressed here in the Hudson Valley, albeit as much narrower features. The deformed sedimentary rocks of the Rondout Valley, Shawangunks, and perhaps the HVB are part of the Valley and Ridge Province. The Hudson Valley is part of the Great Valley, the eastern boundary of the Valley and Ridge. The Hudson Highlands is analogous to the Great Smoky and Blue Ridge Mountains, the Manhattan Prong is the northern tip of the Piedmont, and Long Island and the Jersey Shore are part of the Atlantic Coastal Plain.

Unfortunately, unlike our neighboring states to the south, we don't have a good record of Mississippian and Pennsylvanian rocks exposed here in New York State. This was an interesting time in Earth history where we see evolutionary adaptations in both plants and animals which allowed them to move away from moist environments and eventually colonize all of the terrestrial ecosystems on Earth. We also missed out on the extensive coal deposits from this period of geologic time which may not have been all bad – we also avoided the significant environmental problems seen in many coal mining regions of the Appalachians.

The Mississippian and Pennsylvanian Periods are often collectively referred to as the Carboniferous Period. This was the original name for this span of time and proposed for the famous Coal Measures of north-central England. The Upper and Lower Carboniferous in Europe is roughly equivalent to the Mississippian and Pennsylvanian Periods in North America and the Upper Carboniferous coal deposits in England correspond to the Pennsylvanian coal deposits in the Appalachians. During this period of time, England and North America were in an equatorial climate zone on opposite sides of the old Acadian mountains.

Coal began its life in tropical swamps teeming with plant and animal life. Without much circulation, the water in these swamps became acidic and oxygen-poor over time due to dead vegetation accumulating in the bottom sediments since organic material uses up oxygen and releases acids as it decays. Acidic, oxygen-poor water will preserve organic material and mats of plant fibers, called peat, will accumulate in such environments. As peat is buried and compacted, it eventually drives out water and volatile compounds becoming a sedimentary rock called bituminous coal. This is why coal is referred to as a fossil fuel; it originated from ancient plant remains. If bituminous coal is metamorphosed, it alters to essentially pure carbon and is called anthracite coal – a metamorphic rock. Anthracite is a harder coal, and its higher purity causes it to burn cleaner making it more economically desirable as a fuel. Bituminous coal is more common, but its typically high sulfur levels make it a source of air pollution and acid rain from the formation of sulfur dioxide gas (SO_2) as it's burned. The Pennsylvanian Period sedimentary rocks of Pennsylvania and West Virginia are rich in both bituminous and anthracite coals. The anthracite, however, is limited to belts in the heart of the Appalachians where temperatures and pressures were high enough to metamorphose the bituminous coal into anthracite.

A significant development in the evolution of plants occurred in the latest part of the Devonian Period and is well-represented in the Carboniferous landscape. In the Silurian and Devonian Periods, vascular plants reproduced by spores and required moisture for fertilization to occur. These types of plants are still around, like modern ferns, but they require moist environments to thrive (you'll never see a fern growing in a desert). Near the end of the Devonian, vascular plants developed seeds containing embryonic plants. This was a huge evolutionary advance since seed-bearing plants no longer required moist conditions for fertilization to occur. Seeds can simply be scattered on the winds and plants were now able to move into more arid environments.

Neuropteris fern fossil

The earliest seed-bearing plants are often referred to as gymnosperms (Latin for "naked seed") and are mostly represented today by the conifers – cone-bearing trees like pines, spruces, and hemlocks. Back during the Carboniferous, however, the dominant gymnosperms were the so-called seed ferns. The term seed fern is an oxymoron since true ferns reproduce by spores but the foliage of many tree-like seed ferns did have a fern-frond-like appearance. One commonly-found seed fern fossil in Pennsylvanian coal deposits is Neuropteris which, while it looks very much like a modern fern frond, is not closely related. Seed ferns are extinct today.

Another plant fossil common from Pennsylvanian rocks, and presumably once living in New York as well, is Lepidodendron – a Carboniferous tree that grew up to 50 meters (160 feet) tall. Lepidodendron belongs to a group of seed-bearing plants called lycopods. Lycopods are still around as inconspicuous "ground pine" club mosses growing on Hudson Valley forest floors, but they're a pale shadow of their former stature. Lepidodendron must have been a major component of Carboniferous forests given that some geologists have estimated that their remains compose up to 80% of all the coal deposits in North America.

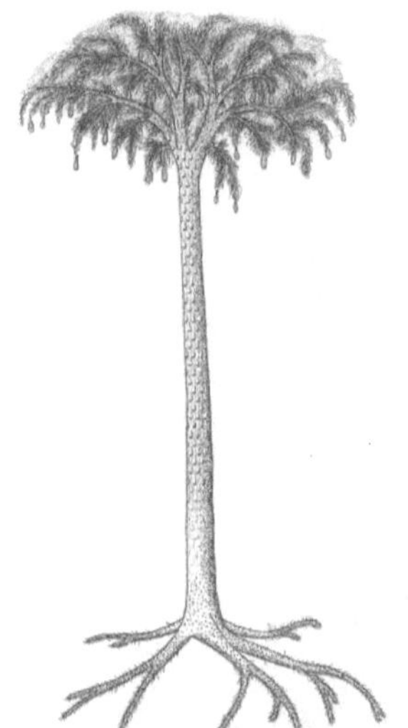

Reconstruction of a Carboniferous Lepidodendron tree and photo of a modern club moss on the Shawangunk Ridge forest floor.

Insects were also important in the Carboniferous forests and some grew to impressive sizes. Giant dragonflies (Meganeura) with wingspans of over 75 centimeters (2.5 feet) were common as well as 30 cm (1 foot) long cockroaches. Insects belong to phylum Arthropoda, the largest of the invertebrate groups in terms of numbers of species. We've already discussed two notable arthropod groups, the trilobites and eurypterids. Modern marine members of the arthropods are animals like shrimp, lobsters, and horseshoe crabs.

Arthropods are distinguished by a segmented body with jointed appendages. The name arthropod literally means "joint foot" in Greek.

Arthropods also have an exoskeleton composed of chitin which is replaced by molting as the animal grows. The possession of an exoskeleton helped arthropods to be the first invertebrates to conquer terrestrial environments in the Silurian Period since it helped protect them from desiccation.

Some arthropods, like modern horseshoe crabs and probably the Silurian eurypterids, developed the ability to come out onto beaches for short periods of time through the development of external book gills, thin leaves of tissue which allow for oxygen gas exchange when kept moist. These evolved into internal book lungs in arachnids like spiders and scorpions.

Insects developed a different type of respiratory system. Small holes in the sides of their thorax and abdomen, called spiracles, admit atmospheric oxygen Tracheal tubes distribute oxygen throughout the body of the insect where it diffuses into all of the body's cells. This method of respiration imposes a size limit on insects based upon possible rates of oxygen diffusion. That's why insects are generally small. So, despite what you may have seen in 1950s science fiction movies, insects are unable to grow to very large sizes because of the unique way they breathe. Why then did insects grow larger than they are today during the Carboniferous Period?

It's believed that the incredible spread of vegetation on Earth during the Carboniferous, as reflected by the abundant coal deposits, led to a depletion of carbon dioxide gas in to the atmosphere and led to an increase in the percentage of atmospheric oxygen. That was due, of course, to photosynthesis which converts carbon dioxide into oxygen during plant respiration. In addition, the widespread Carboniferous tropical swamps kept oxygen levels high because dying plants were quickly submerged in acidic, oxygen-poor waters that kept them from decaying completely, a process which removes oxygen from the atmosphere. Both of these processes contributed to an atmospheric oxygen concentration which was perhaps as high as 35%. By contrast, today's oxygen levels are around 21% (virtually all of the rest is inert nitrogen gas).

These higher oxygen levels would have allowed insects to grow to a larger size. Not Hollywood monster size but you would have had to swat away insects the size of small birds if you were able to travel back to the Carboniferous swamps that once covered our area. The largest insect fossil ever discovered was a dragonfly (*Meganeuropsis*) found in slightly younger Permian rocks of Kansas with a 65 cm (slightly over 2 feet) wingspan! Amphibians during the Carboniferous grew to impressive sizes as well – imagine salamanders several meters in length. Large insects spurred the evolutionary development of large insect eaters.

In the animal kingdom, the big story of the Carboniferous was the evolutionary development of reptiles. Reptiles evolved from amphibians and belong to a larger group of animals called amniotes (mammals are also amniotes). Amniotes have embryos surrounded by special membranes, either in an amniotic egg or within the mother's womb (amniotic sac). The comparison of frog and turtle eggs clearly illustrates the difference. Frogs lay eggs in water and they resemble gelatinous masses. If you take frog eggs out of water, they will dry up and never hatch. When frog eggs do hatch, the larval tadpole stage is wholly aquatic. Over the span of only a few weeks, the tadpoles will start to grow legs (my kids called them "frogpoles") and soon metamorphose into air-breathing frogs – one of the true wonders of nature. Frogs, as amphibians, are thus totally dependent upon water to reproduce.

Reptiles, on the other hand, developed the amniotic egg – an egg with a leathery or hard shell, specialized membranes for gas exchange and waste sequestration, and a large yolk to nourish the developing embryo. The amniotic egg freed reptiles from dependence on water for reproduction and allowed them to conquer the ancient terrestrial world. Reptiles were so evolutionarily successful that the next era of geologic time, the Mesozoic, is often called the "Age of Reptiles" – a time spanning over 180 million years. During this time, reptiles also gave rise to two other important groups, the birds and the mammals (the group from which we obviously evolved).

During the Permian Period, following the Alleghenian Orogeny and assembly of Pangaea, our area had drifted across the equator into the Northern Hemisphere where it has remained to this day. We were, however, still closer to the equator than we are now and within a tropical climate zone. Our area was an uplifted, mountainous terrain dominated by erosion. While we have no sedimentary rock or fossil record from this time in New York, reptiles were certainly walking around through the upland forests.

The assembly of Pangaea throughout the Permian had consequences for both global climate and life on Earth at the time. Clumping all of the continents together into one landmass had an effect on global ocean currents and weather patterns. Some climate models have shown that the central part of Pangaea may have experienced temperatures as high as 45° C (113° F) with variations between winter and summer temperatures ranging by as much as 50° C (122° F). Large areas in the interior of Pangaea were arid deserts due to their distance from the oceans and lack of precipitation. Parts of the supercontinent may have essentially been uninhabitable given these conditions.

The formation of Pangaea also affected life in the oceans. When there are numerous continental blocks on Earth, such as we have today, each continent has its own area of continental shelf with extensive sea life in the relatively shallow waters. When a supercontinent assembles, such as happened in the Permian Period, we reduce the global amount of continental shelf available and increase competition among marine life for the more limited resources. In addition, the southern part of Pangaea extended into far southern latitudes and there's evidence for fairly extensive glaciation in these areas during the Permian. This glaciation would have lowered sea levels exposing large areas of the already reduced areas of continental shelf. We see the environment stresses of this event in the fossil record as many shallow marine organisms became extinct during the Permian Period.

At the very end of the Permian Period, which is also the end of the Paleozoic Era, we see the largest mass extinction in Earth history. That's why there is a fundamental break in the geologic time scale here between the Paleozoic Era ("ancient life") and the Mesozoic Era ("middle life"). Well over 90% (estimates vary) of all species alive at that time abruptly died off and disappeared from the fossil record. No one's quite sure why but there's no lack of hypotheses. Some geologists believe it may have been due to an asteroid impact similar to the one 180 million years later that killed off the dinosaurs. There are problems with this view, however, and in recent years many geologists have implicated a massive volcanic eruption for this mother of all extinction events.

Around 250 million years ago, there was a series of eruptions in what's now Siberia. These eruptions may have continued, on and off, for a million years. While some of the eruptions were explosive, shooting tremendous volumes of ash into the stratosphere and lowering global temperatures for a time, large fissures opening in the crust soon poured out incredible volumes of very liquid basaltic lava. These types of lavas are called flood basalts and are seen in several other geologic provinces around the world but none are as large as those of the Siberian Traps. The word "Traps" is an old term for igneous lava flows (basalts) and comes from a Swedish word meaning "stairs" due to the step-like plateaus formed by the erosion of successive lava flows.

The Siberian Trap lavas cover around 2 million square kilometers (almost 800,000 square miles) of land surface. This much lava could only have been generated by Siberia passing over a large mantle plume during the Late Permian Period. Upwelling heat from the underlying mantle would have melted the lower lithosphere to generate this volume of magma. There are severe environmental effects which would have resulted from such a massive eruption.

With the early eruptions of the Siberian Traps, large amounts of dust were emitted into the atmosphere blocking sunlight and reducing global temperatures for a time, perhaps increasing the amount of glaciation and lowering sea levels some more. As flood basalts started erupting, the amount of dust is reduced but now copious amounts of volcanic gases are emitted. Gases like sulfur dioxide (SO_2) will mix with atmospheric water creating sulfuric acid (H_2SO_4) and a very acidic rainfall. Other gases like chlorine and fluorine emitted during volcanic eruptions are highly reactive elements that can destroy ozone in the Earth's stratosphere. The ozone layer protects organisms on the surface of the Earth from damaging ultraviolet radiation.

The most important gas emitted during eruptions, however, is carbon dioxide (CO_2). Carbon dioxide is a greenhouse gas which very effectively traps heat in the atmosphere. Over the tens to hundreds of thousands of years during which the Siberian Traps erupted, global temperatures would have risen significantly from the increased carbon dioxide gas. Studies of carbon isotopes in sediments from this time, however, indicate something even more significant. Large amounts of methane (CH_4) gas may have also been released during this time. Methane is about ten times more effective as a greenhouse gas than is carbon dioxide. Since volcanoes typically don't emit large quantities of methane, where did it come from?

Over the past few decades, geologists studying the ocean floor have discovered that there are immense deposits of methane trapped within sediments on the continental shelves. This methane originates from the bacterial breakdown of organic material in shallow seafloor sediments where the high water pressure and low temperature (near $0°$ C) cause the gas to become trapped in ice crystals called methane hydrates (also called methane clathrates). Warming of the atmosphere and oceans during the Permian, along with lower sea levels, may have caused these methane hydrates to volatilize with a large burp of methane into the atmosphere which would greatly increase the rate of global temperature rise. These high temperatures, by the end of the Permian, resulted in a desert-like arid landscape throughout much of the interior of Pangaea and widespread anoxia in the Permian Oceans. Life at the time couldn't cope quickly enough to these widespread climatic changes and was hit hard. Life on Earth almost came to an end during this time. We're all lucky to be here today.

Life, always resilient, slowly bounced back from the massive Permian extinction. It was, however, a very different world as reflected in the fundamental boundary at this time between the Paleozoic ("ancient life") and Mesozoic ("middle life") Eras. The subsequent Triassic Period begins the "Age of the Dinosaurs" and there's actually a slice of time from this period preserved here in the Hudson Valley as we'll see in the next chapter.

11
Rifting

The earth is broken up, the earth is split asunder, the earth is thoroughly shaken.

Isaiah 24:19 (NIV)

*I*n the above quotation, the Biblical prophet Isaiah is clearly describing an earthquake during the last days. Those listening to Isaiah during the mid-8th century BCE would have been very familiar with this imagery as a large earthquake occurred in Judea a few years earlier during the reign of King Uzziah (see Amos 1:1 and Zechariah 14:5). Earthquakes were familiar to the ancient Hebrews since they lived along the Jordan Rift Valley – a place where the continental crust is undergoing tensile stresses and pulling apart. As continental crust pulls apart, movement of rock along brittle faults will generate large earthquakes. The Jordan Rift is the northern part of a fracture zone that begins down in Africa as the East African Rift.

The African tectonic plate is about twice as large as the continent of Africa and includes the eastern half of the southern Atlantic Ocean and a large part of the western Indian Ocean. There is, apparently, a mantle plume beneath northeastern Africa resulting in uplift, tensile stresses within the crust, and rifting of the continent between the Horn of Somalia (Somalian subplate) and the rest of the continent (Nubian subplate). This is known as the East African Rift System and cuts across 3,000 kilometers (over 1,800 miles) of Africa from the Red Sea in the north to the Zambezi River in the south. Note that there are two parallel rift systems – an eastern and a western valley. This modern-day rifting event is very similar to what occurred 200 million years ago in what's now the Hudson Valley.

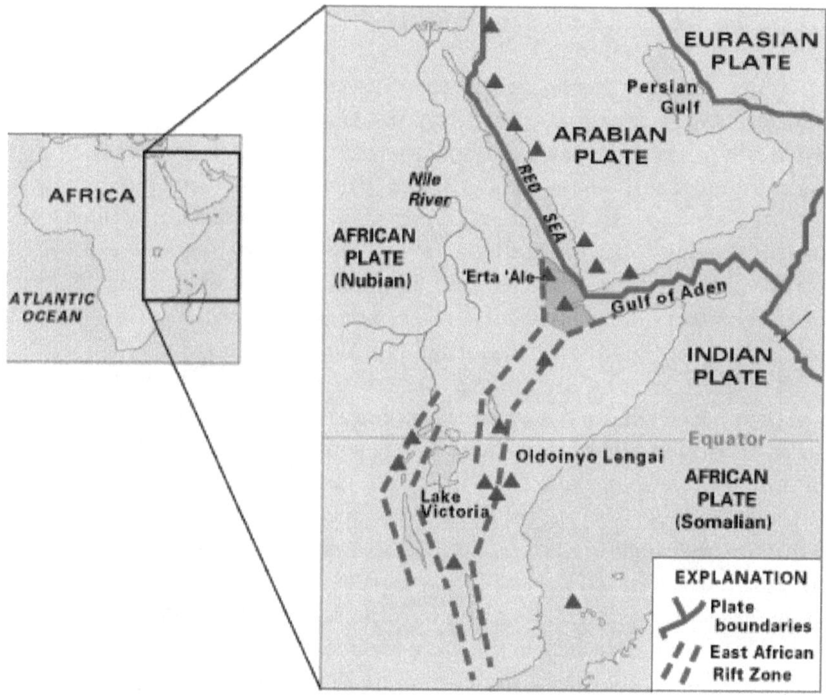

East African rift system. Triangles are volcanoes

At the start of the Mesozoic Era, Pangaea had formed from the amalgamation of several blocks of continental crust which collided and accreted to form a supercontinent. Supercontinents, however, are inherently unstable. The thick continental crust acts as a thermal blanket over the underlying mantle allowing heat to build up at the base of the lithosphere. As with Rodinia, over 500 million years earlier, heat built up beneath the supercontinent of Pangaea, possibly due to the development of a mantle plume, which initiated rifting and eventually began to break it apart into the present-day continents. The breakup of Pangaea was a complex process proceeding in several distinct stages.

The first stage was a rifting event, extending in time from the Triassic into the Jurassic Period, which eventually broke Pangaea into two large continents – Laurasia in the North and Gondwana in the south. This is reflected in our part of the world by the rifting of Africa from eastern North America eventually forming the present-day Atlantic Ocean. This event directly affected the southern part of the Hudson Valley as we'll soon see. Later stages of rifting, throughout the Mesozoic Era and into the early Cenozoic Era, first broke apart Gondwana into Africa, South America, Australia, and Antarctica and then broke apart Laurasia into North America, Greenland, and Eurasia to eventually form the continents we're all familiar with today.

When Pangaea began to rift, the continental crust was subjected to tensile stresses acting to pull it apart. At depth, where rocks behave plastically, these stresses stretched and thinned the crust like soft taffy. Near the surface, however, rocks are brittle and faults formed in response to these stresses. In Chapter 8, we discussed the Hudson Valley fold-thrust belt and introduced thrust faults. These faults formed in areas of compressive stresses where crustal shortening is occurring due to continental collisions. In areas of continental extension, however, a different type of fault, called normal faults, will develop.

To illustrate the difference between these two fault types, it's necessary to introduce some terminology. When looking at a fault in cross-section as it dips into the subsurface, we can see that there are two blocks of rock on either side of the fault plane. The block below the fault is traditionally called the footwall block and the one above is called the hanging wall block. Faults, being fractures in the rock, are often conduits for groundwater flow, and so we find that mineral deposits are sometimes precipitated along them. Mine tunnels, following the mineral deposits, would then follow the fault surface. If you can imagine standing in such a mine, the block under your feet would logically be called the footwall block and the one over your head the hanging wall.

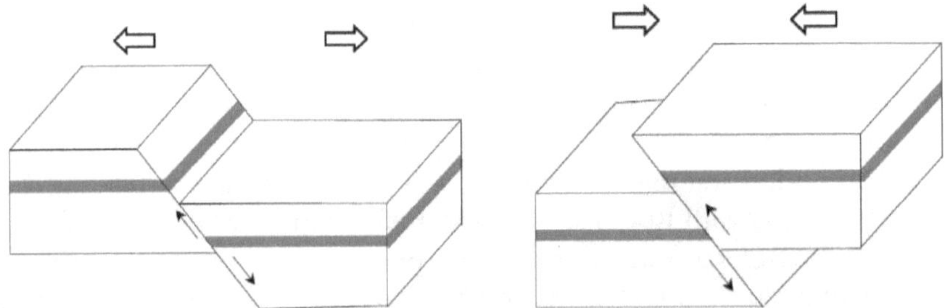

Normal fault at left (hanging wall down) and reverse fault at right (hanging wall up). Normal faults accommodate crustal extension and reverse faults accommodate crustal compression

If you were to pull apart a piece of crust cut through by a fault, you can see that the hanging wall block will slide down the ramp of the footwall block as the crust lengthens. This is called a normal fault (think of it as the "normal" way a block will slide down a ramp). Conversely, if you were to compress the same piece of crust, you can see that the hanging wall block will move up the ramp of the footwall block as the crust shortens. This is called a reverse fault (the reverse of "normal" movement). Thrust faults are a special type of reverse fault, typical of fold-thrust belts, where the fault plane has a dip angle of less than 45° (generally less than 30°). It should now be obvious why reverse or thrust faults are associated with continental collision since they accommodate shortening of the crust and why normal faults are associated with continental rifting since they accommodate lengthening.

Rift zones, being areas of continental extension, are therefore associated with normal faults. As blocks of rock drop downward, they form a central rift valley and generate shallow crustal earthquakes. Rift valleys, being low areas, tend to accumulate clastic sediments like pebbles and sand from the weathering of higher areas on either side of the valley. Over time, water may accumulate in these valleys and form long, linear lakes. Volcanic activity may also be present in rift valleys since the crust is thinner and, as the crust thins, there's less pressure on the underlying mantle which will initiate partial melting to form magma. This is called decompression melting since the melting temperature of rock is proportional to the pressure it's under and rifting will reduce the pressure enough to cross that melting point.

All of these features are seen in modern rift zones like the East African Rift. This system, as mentioned previously, has formed two parallel valleys in eastern Africa and is a seismically active area. Earthquakes are generated along the normal faults on either side of these valleys since the rift is pulling apart at a rate of several millimeters each year. In 2006, for example, an impressive magnitude 7.4 earthquake occurred in the southern part of the rift zone in Mozambique but smaller quakes are much more common.

The relatively rapid accumulation of clastic sediments in the northeastern part of the Rift Valley is also significant because they've preserved some very interesting fossils. These fossils are geologically young, still preserved in unconsolidated sediments instead of sedimentary rock, but important because they represent our earliest ancestors. A number of different hominid species evolved in eastern Africa over the past few million years and are found buried in the Rift Valley sediments. Rift volcanism has also deposited ash layers allowing for the age of some of these fossil remains to be constrained by radiometric dating.

One example of such a fossil find is Lucy, discovered in 1974 by paleoanthropologist Donald Johanson at a site called Hadar in Ethiopia.

While only about 40% of her skeleton was recovered, Lucy is significant because she shows clear evidence of being a relatively small-brained, petite, bipedal hominid who was walking around the East African Rift area about 3.2 million years ago. Nicknamed after its discoverers heard the Beatle's song Lucy in the Sky with Diamonds on the radio she's been assigned to the species Australopithecus afarensis. While a completely different species from us, Australopithecenes are thought to be ancestral to our own genus of Homo. Hundreds of other fossil fragments of Australopithecenes have been recovered from eastern Africa as well, and in 1978, Mary Leakey, wife of the famed paleoanthropologist Louis Leakey and a distinguished scientist in her own right, discovered footprints of three Australopithecenes (a family, perhaps?) in 3.7 million-year-old volcanic ash deposits at Laetoli, Tanzania. If it wasn't for the East African Rift, we'd have a much poorer picture of our evolutionary history.

A number of large linear lakes are present within the low-lying areas of the East African Rift. Lake Tanganyika, for example, which lies between Tanzania and the Congo is an example found in the western arm of the rift. Lake Tanganyika is over 650 kilometers (400 miles) long but only about 50 km (30 miles) wide. It's some 1,500 meters (almost 5,000 feet) at its deepest point, with an additional 4,500 kilometers (2.8 miles) of sediment before hitting bedrock. It's the second largest and second deepest freshwater lake in the world. The largest and deepest is Lake Baikal in Siberia, another linear rift valley lake. The eastern arm of the East African Rift, with its hot, arid climate, is characterized more by alkaline lakes with high evaporation rates (these are often called soda lakes since they have high concentrations sodium bicarbonate).

Finally, volcanoes are also present in the East African Rift. Mount Kilimanjaro, in Tanzania, is an example of a presently inactive volcano. One of its peaks, Uhuru, is the continent's highest at just under 6,000 meters (20,000 feet). Although Kilimanjaro hasn't erupted in historic times, it does actively emit volcanic gases through several fumaroles near the summit and there are indications of a magma chamber 400 meters (1,300 feet) below the old crater. It's not extinct; it's simply dormant for now. Other volcanoes, however, are currently active within the East African Rift Valley.

All of the features that are today found in eastern Africa were once present in eastern North America when Pangaea began to rift apart. While the main rift valley is now gone and replaced by the ever-widening Atlantic Ocean, parallel rift valleys that formed during this time are preserved in a number of locations along the East Coast. This is analogous to what we would see if one of the arms of the East African

Rift broke through to form an ocean basin – the other arm would be preserved as a failed rift basin on the continent.

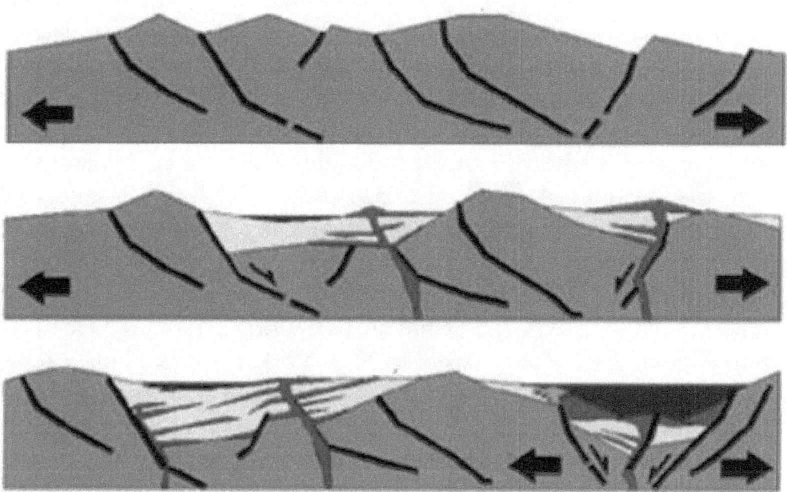

Top: Pangaea begins rifting from crustal extension with the formation of large-scale normal faults. Middle: Rift valleys form with lakes, the influx of sediments, and volcanic activity. Bottom: One rift valley becomes the Atlantic Ocean while others are left behind like the Newark Rift in southeastern New York and northeastern New Jersey.

These aborted rifts in eastern North America can be thought of as stretch marks in the crust. The Newark Basin is the most important of these rift basins for Hudson Valley geology and extends from northern New Jersey into Rockland County. In New York, the basin is bounded on the north and west by the Ramapo Fault – an abrupt contact between the Hudson Highlands and Newark Lowlands also called the border fault. On the east, the basin is bordered by the Hudson River and rocks of the Manhattan Prong are found on the opposite side of the river.

The Newark Basin is contained within a structural feature in the crust called a half graben. Graben is German for "ditch" and it represents a down-dropped block of crust bordered by normal faults. The Newark Basin is a half graben because the western edge has a pronounced border fault, the Ramapo Fault, while the eastern edge is a fold hinge where the rocks and sediments have simply been bent downward. In the deeper western part of the basin, well over 3 kilometers (10,000 feet) of sediment accumulated during the Late Triassic Period as Pangaea was rifting apart a short distance to the east.

There are some notable sedimentary rock formations within the

Newark Basin starting in the Late Triassic and extending into Early Jurassic Periods. The oldest is the Stockton Formation, a coarse clastic deposit represented primarily by buff to reddish arkose. Arkose is a type of sandstone containing significant amounts of feldspar minerals along with quartz sand grains. Feldspars are a group of minerals common in rocks like granite and granitic gneiss, both of which are present within the Hudson Highlands. They are relatively unstable minerals on the Earth's surface; in the presence of slightly acidic water, feldspars will quickly weather into clay minerals. Because of this, arkosic sandstones are typically formed in more arid climates relatively close to the source region of their sediments. The source region would in this case be the Hudson Highlands adjacent to the rift valley in Triassic times.

The Stockton is exposed at the surface along a narrow belt on the eastern edge of the Newark Basin. In New York, this would place it next to the Hudson River and below the Palisades Sill (which we'll discuss shortly). In the Nyack area, a number of abandoned quarries are found in the Stockton Formation where it was mined for brownstone over 100 years ago. The sedimentary rocks within the Newark Basin are brownish to reddish due to an abundance of iron present within the sediments. In the terrestrial environment in which these sediments accumulated, atmospheric oxygen was readily available to oxidize the iron into minerals like rusty-red hematite (Fe2O3) and yellow-brown limonite. Strictly-speaking, limonite is not a mineral but rather a mixture of iron oxide (FeO) and iron hydroxide (FeOH) minerals.

Brownstone was an important building stone in the 19th century and brownstone row houses can be found in cities throughout the northeastern United States. In New York City, they're especially common in some areas of Brooklyn. Brownstones fell out of favor at the turn of the 20th century, around the time Portland cement became widely available, but have recently undergone a revival in popularity in some urban areas. One of the problems with brownstone as a building material is that sandstone is porous and water seeping into the pores will eventually cause spalling of surface layers after being exposed to years of freeze-thaw cycles during the harsh northeastern winters.

Much of the brownstone in New York City came from quarries in the Triassic arkosic sandstones of the Newark Rift. Nutley, for example, just 16 kilometers (10 miles) west of Manhattan in New Jersey, had a number of working quarries in the 1800s. The largest and best known quarry in the northeast, however, was the immense Portland Brownstone Quarry on the banks of the Connecticut River in central Connecticut. The Hartford Basin, running from Connecticut northward into Massachusetts, is very similar geologically to the Newark Basin but the

Portland Quarry had two things going for it – first, its location on a major river allowed schooners to ship the brownstone to cities like Hartford, New York City, Boston, and Philadelphia and, second, the Triassic sandstones in the Portland area were horizontally bedded and very easy to effectively quarry. Most operations shut down in the 1930s when a major flood inundated the quarries. They are now preserved as a city park although limited quarrying still occurs in the nearby area.

Brownstones in Harlem, New York City

Brownstone quarries in northeastern rift basins are important not only for their historical significance but also for the fact that many dinosaur trackways have been recovered from sandstone slabs during the heyday of quarrying operations. Unfortunately, while tracks are relatively abundant, virtually no fossil bone material has been recovered in these environments. Given this, it's sometimes difficult to match up a track to the exact species of dinosaur that created it on some long-ago Triassic afternoon. Oftentimes, tracks are assigned to what are called ichnogenera. Ichnofossils (*ichnos* is Greek for "trace" or "track") are traces of an organism's activity like tracks, trails, or burrows. Three common ichnogenera assigned to Triassic tracks are *Grallator*, *Anchisauripus*, and *Eubrontes* for small, medium, and large three-toed, bipedal dinosaur tracks respectively. We can only guess as to what exact species of dinosaur created these different tracks. We'll return to this shortly.

Returning to the Newark Basin in the Hudson Valley area, the next youngest rock unit above the Stockton is the Passaic Formation. The Passaic is sometimes referred to as the Brunswick Formation in the older literature. Most of the Newark Basin in Rockland County, New York is underlain by this rock unit which consists primarily of dark-colored siltstones and shales alternating with reddish shales. This is the most widespread unit in the Newark Basin with a thickness of several kilometers.

The Passaic Formation, along with another unit called the Lockatong Formation exposed to the south in New Jersey, are cyclical deposits representing the repeated expansion, high stand, and drying up of lake beds within the Newark Rift Valley. Geologists refer to lake-deposited sediments as lacustrine deposits and these cycles in the Newark Basin have been called Van Houghten cycles after the Princeton University geologist who first described them. At times when these lakes covered virtually the entire basin, the sediments preserved a variety of fish and other freshwater marine fossils such as crustaceans and mollusks. Some may have rivaled the East African Rift Valley lakes in size. At other times, the lakes dwindled in size, eventually dried up, and we find numerous Triassic plant fossils and dinosaur tracks in the sediments. These Van Houghten cycles repeated over tens of thousands of years and represented climatic changes over time. The likely reason for these climatic changes are cyclical astronomical changes in the Earth's tilt, axial wobble, and orbital shape. These are called Milankovich cycles and will be described in more detail in Chapter 13 when we discuss glaciation in the Hudson Valley. Changes due to tectonic forces associated with the breakup of Pangaea and episodic movement on the basin's border faults certainly played a role in lake-level changes as well.

Another Newark Basin rock unit in southeastern New York is the Hammer Creek Conglomerate. This rock unit is found primarily on the western edge of the basin next to the Ramapo Fault and adjacent Hudson Highlands and is probably best considered as simply a conglomeritic facies of the Passaic Formation (remember sedimentary facies?). These rocks represent what's known as a fanglomerate, a conglomerate formed within an alluvial fan. Alluvial fans are structures common today in arid areas where flat-lying basins exist next to uplifted mountains such as Death Valley in California. Death Valley is actually a pretty good analogy to the Newark Basin during its more arid phases since it's also a fault-bounded, down-dropped basin formed by crustal extension. When it rains, water quickly flows down out of the mountains through narrow canyons, picking up tremendous amounts of clastic sediments. As the water reaches the flat valley floor, it spreads out, loses velocity, and drops sediment. Coarse sediments are dropped first near the mouth of the canyon and finer one are carried further out into the basin. These deposits build up over time to form an apron of sediment called an alluvial fan.

Along the Ramapo border fault, the Hudson Highlands were being uplifted as the Newark Basin was being down-dropped. Streams flowing off the higher Hudson Highlands formed alluvial fans as they flowed out onto the flat Newark Lowlands. These alluvial fans are preserved today as deposits of the Hammer Creek Conglomerate which contain rounded cobbles of gneiss and other igneous and metamorphic rocks from the Hudson Highlands clearly attesting to their source region.

Large cobbles and pebbles in the Hammer Creek Conglomerate off Route 202 in Rockland County

We now get to one of the most famous rocks of the Newark Basin, the Palisades Sill. From Giovanni da Verrazzano's first sighting in 1524 to the modern day, the cliffs of the Palisades have been an important landmark along the shores of the southern Hudson River. The name, of course, comes from its resemblance to a defensive wooden fence composed of sharpened, vertically-arranged tree trunks. The vertical fractures in the cliffs of the Palisades are due to the cooling and contraction of this mass of igneous rock forming what geologists call columnar joints, the same feature seen in places like Devil's Tower in Wyoming.

Palisades Sill from the Hudson River

The Palisades are composed of the igneous rock diabase. The basic difference between diabase and basalt is that diabase is a bit more coarse-grained because it cools a little slower in the shallow crust rather than quickly at the surface. Slower cooling allows for larger crystal growth. During the formation of the Newark Basin, decompression melting beneath the thinner crust of the rift formed molten rock which intruded up into the sedimentary rock of the basin where it forced its way laterally between horizontal layers of the Stockton, Lockatong, and Passaic Formations. Intrusive, horizontal, tabular bodies of igneous rock formed this way are called sills. The Palisades Sill has been exposed by erosion in a relatively narrow band from Staten Island northward along the western shore of the Hudson River to the Haverstraw area in Rockland County where it hooks a few kilometers to the west. In north-central New Jersey, and at one location in southeastern New York (Ladentown), molten rock from the sill actually found its way upward through feeder dikes to form surface lava flows. Today, these lava flows form the Watchung Mountains in New Jersey. The reason lava flows can form mountains is because the lavas are a bit more resistant to erosion than the surrounding soft sedimentary rocks of the Newark Basin and thus stand in relief after almost 200 million years of weathering.

Radiometric dating of the Palisades Sill provides an age for this volcanism of around 190 million years, which places the intrusion into the Early Jurassic Period of geologic time. Geochemical studies of the Palisades Sill reveal that there were two or more separate injections of magma to form this structure. While the cliffs of the Palisades are impressive, up to 165 meters (540 feet) in height, the sill dips downward toward the west by 10-15° in the subsurface where it averages around 300 meters (1,000 feet) in thickness.

There are a number of interesting geologic features that can be examined in the area around the Palisades Sill. Along the upper and lower margins of the sill, there is a zone of contact metamorphism where the surrounding sedimentary rocks were altered by the high heat of the molten magma intrusion. These metamorphic rocks are called hornfels. Near the upper margin of the sill, it's sometime possible to find inclusions of the overlying sedimentary rock incorporated into the igneous rock of the sill. These are called xenoliths, from the Greek for "stranger rock," due to their out-of-place nature. They represent pieces of the host rock which broke off, fell into the magma, and were able to survive the high heat without totally melting away. Near the bottom of the sill is a 3 meter (10 feet) thick zone enriched in the mineral olivine. Olivine is the first mineral to crystallize when a magma is cooling and this zone represents an accumulation of minerals that cooled and settled to the bottom of the sill before the rest of the magma cooled and solidified.

The Palisades Sill is easily viewed from across the river in Manhattan and can also be seen closer up from your car on the New Jersey side of the Lincoln Tunnel and George Washington Bridge. For a more leisurely look, other than an early-morning traffic jam on the western bridge and tunnel approaches, one excellent place to spend some time is the Palisades Interstate Park. Between Fort Lee on the western side of the George Washington Bridge and State Line Lookout at the New Jersey/New York border, the park offers scenic drives and hiking trails along both the top and bottom of the Palisades. A number of state parks in New York also allow hiking to view different parts of the Palisades ridge including Tallman Mountain, Blauvelt, Rockland Lake, Hook Mountain, and High Tor State Parks. Most people have no difficulty identifying the igneous rocks of the Palisades Sill when encountering them in the field.

From Manhattan, the Palisades look relatively pristine today, but this was not always the case. From the earliest days of European settlement, the rocks of the Palisades, called trap rock, were quarried for use as building stones, ship ballast, and crushed rock. Concerned over the ever-expanding plans to quarry the Palisades, efforts were made in the late 18th and early 19th century to protect and preserve this unique landscape. A number of wealthy New York City philanthropists made possible the purchase of land along the New Jersey side of the Hudson River and the creation of the Palisades Interstate Park Commission thus preserving the views from their Upper West Side Manhattan homes. One place the diabase of the Palisades Sill is still mined today is at Tilcon's Haverstraw Stone Quarry in Rockland County, New York.

Remember the Shawangunk Ridge? In Chapter 6 we discussed how fractures and faults within the Silurian Period conglomerates contained minor deposits of lead-zinc minerals that were historically mined from Ellenville southward. While the fractures and faults in the Shawangunks formed during the Alleghenian Orogeny, much later than the conglomerates themselves were deposited, the minerals precipitated even more recently. Radiometric dating of mineral inclusions within the vein-filling quartz suggest an age range in the Early to Middle Jurassic Period – a time when the Newark Basin was forming.

If these radiometric ages are accurate, it indicates that Shawangunk faults were reactivated during the rifting of Pangaea even though the Ridge is located tens of kilometers to the west of the Mesozoic rift valley. Reactivation of preexisting faults is very common in deformed areas, it's far easier to move rock on a fault that's already there than it is to create a fresh fracture in intact rock. Presumably, stresses from the breakup of Pangaea extended far enough westward to initiate some movement, and fluid flowing through these fractures precipitated quartz along with the lead-zinc minerals like galena (PbS) and sphalerite (ZnS).

This illustrates an important point in geology – even small areas can have very long and complicated histories with different parts of a single rock outcrop forming at separate times for unrelated reasons over a span of hundreds of millions of years. With the Shawangunk Conglomerate, for example, the quartz likely formed over a billion years ago in the Grenville Mountains, the pebbles formed during the Silurian Period in a fluvial environment, the rocks were deformed by folding and faulting during the Alleghenian Orogeny, these fractures were intruded by mineral-precipitating fluids during the Jurassic Period, and everything was later exposed for our study by erosion and glaciation in recent times.

Let's return to the big picture. While Pangaea was rifting apart, the big story in the history of life was the evolution of reptiles. While there were many groups of reptiles living during the Mesozoic Era, the best known are obviously the dinosaurs. At the very end of the Paleozoic Era, a group of reptiles called archosaurs arose which gave rise to the dinosaurs in the early Triassic Period. While dinosaurs are extinct, archosaurs are still around today in the form of crocodilians and birds (birds are the evolutionary descendants of dinosaurs). Dinosaurs are traditionally classified into two taxonomic orders based upon their hip joints - the saurischian, or lizard-hipped dinosaurs, and the ornithischian, or bird-hipped dinosaurs. These names are misleading; birds actually evolved from the lizard-hipped dinosaurs!

There are two main groups of saurischians, the theropods and the sauropods. Theropods are bipedal and the only carnivorous dinosaurs. Examples are the infamous Tyrannosaurus rex and the Velociraptor of Jurassic Park fame (notwithstanding some inaccuracies in the film's portrayal of these very real animals). Sauropods were quadrupedal (four-legged), herbivorous, and the largest of the dinosaurs. Well known sauropods include Apatosaurus and Diplodocus. All of the other types of dinosaurs fall into several different subgroups of the Ornithischians.

The earliest dinosaurs that evolved from the archosaurs in the Triassic Period were theropods and we know that some of them walked around in what's now the Hudson Valley around 200 million years ago. In 1972, dinosaur tracks were discovered in a reddish mudstone from the Newark Rift along the Hudson River just north of Nyack in Rockland County. The slab containing these tracks is today on display in the lobby of the New York State Museum in Albany. These three-toed dinosaur tracks were clearly made by a small theropod similar to *Coelophysis*, a common dinosaur of the same age and known from numerous fossils out west (especially a place called Ghost Ranch near Abiquiu, New Mexico).

Coelophysis was an ostrich-like, carnivorous dinosaur that stood about 1.5 meters (5 feet) high and weighing a bit over 45 kilograms (100 pounds). While small, its sharp teeth and claws along with its speed and agility would have made it a fearsome predator in the Late Triassic world. There is some evidence that, unlike the solitary T. rex which arose later in the Mesozoic Era, *Coelophysis* may have hunted in small packs. There's also increasing evidence that many of the theropods like *Coelophysis* may have been endothermic or warm-blooded. This is not such a stretch when considering that the warm-blooded birds evolved from such dinosaurs.

Coelophysis bauri reconstruction

While we don't have a great fossil record of dinosaurs here in New

York State, there are numerous ichnofossil trackways representing a diverse group of theropod dinosaurs in the New Jersey part of the Newark Basin as well as in the similar-aged sediments in the Hartford Basin of central Connecticut and Massachusetts. These dinosaurs were presumably all walking around New York State as well and it's certainly possible that we'll one day find more evidence of their existence in the rocks of Rockland County (although it's harder to do now since much of the area is paved-over suburbia).

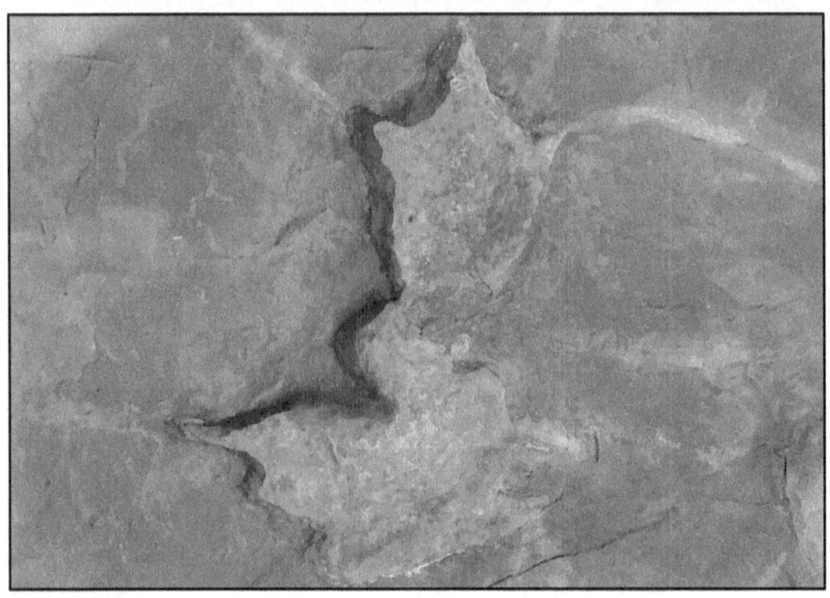

Theropod dinosaur track in Triassic shale from Rockland County

What was the environment like during the Late Triassic Period? The climate in what's now the Hudson Valley was subtropical and typically arid. Deciduous trees and grasses had not yet evolved and the landscape was covered with conifers, cycads, ginkos, ferns, and horsetails (Equisetum). The meat-eating theropod dinosaurs ranging across the area left footprints along muddy lakeshores as they searched for food while a few early semi-bipedal herbivorous dinosaurs (Anchisaurus) grazed on the tough vegetation.

Other reptiles also existed in this environment. Numerous specimens of 30 centimeter (1 foot) long lizard-like reptiles (*Hypsognathus*) and 8 meter (26 feet) long crocodile-like phytosaurs (*Rutiodon*) have been recovered from the Newark Basin. Lobe-finned fish (*Diplurus*) and amphibians (*Metoposaurus*) also appeared on muddy tidal flats next to expansive rift-valley lakes. Numerous species of primitive bony fish swam in these lakes which also supported a rich variety of fresh-water invertebrates. The buzzing of insects would have filled the air during the hot afternoons as we stand on the brink of the Jurassic Period – a time when the Earth was truly ruled by dinosaurs.

As we leave the Late Triassic/Early Jurassic landscape, Pangaea has begun the process of breaking apart into the continents we know today. No longer do we speak of Laurentia, now we can refer to our block of crust as North America. Until the great Pleistocene Ice Age, however, most of the Hudson Valley Region preserves virtually no sedimentary record for almost 190 million years.

12
The Long Gap

Time, geologic time, looks out at us from the rocks as no other objects in the landscape. Geologic time! How the striking of the great clock, whose hours are millions of years, reverberates out of the abyss of the past! Mountains fall, and the foundations of the earth shift, as it beats out the moments of terrestrial history. Rocks have literally come down to us from a foreworld.

John Burroughs, *Under the Apple Tree*

When we left off in the previous chapter, Pangaea was rifting apart during the Late Triassic/Early Jurassic Periods. From that time to the present day, the Atlantic Ocean has been growing ever wider as North and South America drift apart from Europe and Africa at a rate of around 2.5 centimeters (1 inch) each year.

How did an ancient rift valley turn into a modern ocean basin? As rifting continued from the Triassic into the Jurassic Period, the continental crust beneath several of the rift valleys in what's now eastern North America stretched and thinned until they eventually linked up into a through-going fracture between North America and Africa. At that point, magma formed in the underlying mantle by decompression melting, rose through the crust, and solidified into new rock. This new crust, derived from partially-melted mantle material, was basaltic in composition. Seafloor crust under all the world's oceans is composed of the igneous rock basalt created by this process of sea-floor spreading.

The Mid-Atlantic Ridge is a classic example of a sea-floor spreading ridge. It was first discovered in 1872 and extensively studied after the use of SONAR technology was applied to oceanographic research following World War II. This ridge is part of a larger mountain chain almost 60,000 kilometers (over 37,000 miles) in length, running through the middle of the world's ocean basins. Unlike terrestrial mountains, however, mid-ocean ridges have gentle slopes since they're very broad compared to their height above the seafloor. The Mid-Atlantic Ridge averages 2,000 kilometers (1,240 miles) wide and reaches elevations of 2.5 kilometers (8,200 feet) above the seafloor. Even with these elevations, the top of the ridge is still another 2.5 kilometers below the wave-tossed surface of the Atlantic.

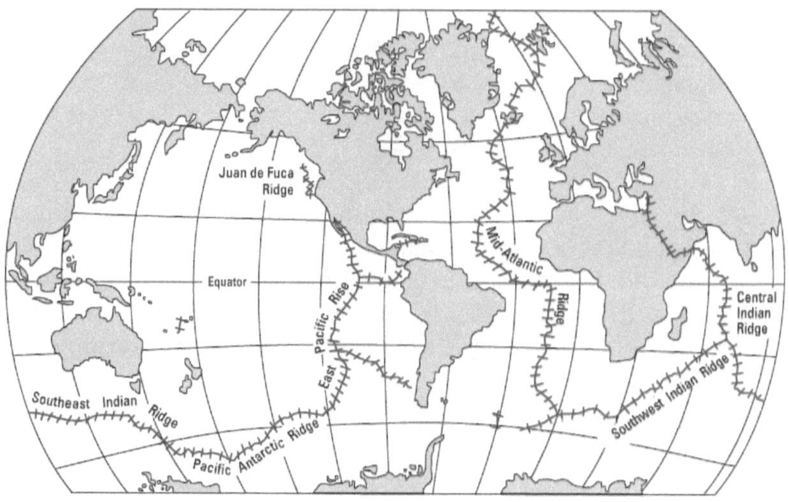

Worldwide mid-ocean ridge system

One might reasonably ask why a rift in the seafloor crust is expressed as a mountain range. After all, rifts in continental crust form rift valleys. The answer is that the seafloor crust is much thinner than continental crust. On average, oceanic crust is about 7 kilometers (4.3 miles) in thickness compared to an average thickness of continental crust of 35 kilometers (22 miles). Continental crust can also be up to twice that thickness under large mountain ranges like the Himalaya. The oceanic crust, being thinner, is more easily heated by the hot, rising mantle material found beneath a spreading ridge. When the crust is hotter, it will literally "float" higher on the mantle than when it's cool. As crust moves away from the spreading ridge, and new crust forms there to replace it, the older crust cools and sinks. Thus a ridge exists over the spreading center.

In cross-section, oceanic crust shows some interesting details. At the very top are seafloor sediments which thicken as you move away from the ridge. This is because crust is youngest at the ridge and hasn't had much time to accumulate sediments. Below the sediments is the basaltic rock of the oceanic crust. This rock, however, shows different structural details with depth. Near the surface of the seafloor, the oceanic crust is composed of pillow basalts. These are rounded masses of basalt which form when lava is extruded into cold seawater and quickly cools. Earlier, in Chapter 4, we discussed Stark's Knob which is an Ordovician Period pillow basalt exposed along the Hudson River at Schuylerville.

Below the pillow basalts are vertical, tabular bodies of basalt called sheeted dikes. These formed when magma was extruded upward through vertical fractures as the seafloor was pulled apart by the tensile forces of seafloor spreading. Below the sheeted dikes is a fossilized magma chamber composed of the igneous rock gabbro. Gabbro is exactly the same composition as basalt, the only difference is that it's coarser-grained since it was buried more deeply and it had time to grow larger crystals as it cooled and solidified.

Below the gabbro is peridotite, the rock which comprises the Earth's mantle. Peridotite is typically greenish in color since it's primarily composed of the olive-green mineral olivine. Olivine is a simple mineral, it's a silicate with iron and/or magnesium mixed into its composition – $(Fe,Mg)_2SiO_4$. If you take mantle peridotite, and heat it up until it just starts to melt, the partial melting will form a magma with the composition of basalt.

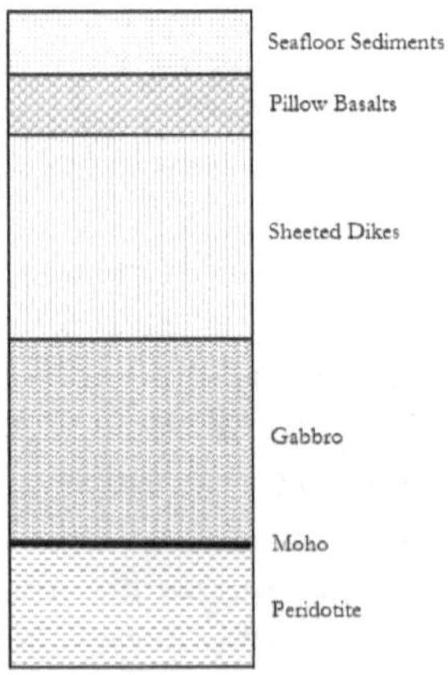

Structure of the oceanic crust and mantle

It's easy to see how this structure developed within the seafloor given that all seafloor crust develops at mid-ocean ridges. What truly amazed geologists when this structure was revealed from the deep-sea drilling programs starting in the 1960s was that geologists had been finding these rock types and structures on land for decades but hadn't realized they were looking at pieces of the seafloor crust. Who would even expect seafloor crust on land, especially in mountainous areas? Remember the Staten Island serpentinite we discussed in Chapter 5? This was a piece of seafloor crust caught up in the Taconic Island Arc collision during the Ordovician Period and permanently sutured into the North American crust.

When slices of oceanic crust are thrust up onto the continents they are called ophiolites. The term comes from the Greek words for serpent (*ophis*) and stone (*lithos*) because of the mottled green color of the metamorphic rock serpentinite which typically forms in ophiolites from the hydrothermal alteration of basalt. The process by which these ophiolites are emplaced is called obduction since it's essentially the opposite of the process of subduction. Ophiolites are found throughout the Appalachians and were emplaced during the three major Paleozoic Era collisional mountain building events.

During the time the Atlantic Ocean was forming, New York State remained mostly above sea level as an area of erosion as it is today. That's why we have no rocks preserved for almost 100 million years through much of the Mesozoic Era. The only exception to this is along the coastal plain found on both Staten Island and Long Island. The Coastal Plain was created after Mesozoic rifting formed the Atlantic margin of North America, and gently slopes downward until it eventually dives below sea level becoming the continental shelf. The coastal plain and continental shelf are collectively about 300 kilometers (186 miles) wide. Along the northern section of the East Coast, as in New England, there is essentially no coastal plain and a very wide continental shelf. From New Jersey southward, however, the coastal plain is fairly wide and the continental shelf slightly narrower to compensate.

Keep in mind that the present-day shoreline is an ephemeral thing. A slight rise in sea level will flood large areas of the Atlantic and Gulf coastal plains. A slight drop in sea level will expose large areas of the continental shelf and move the shoreline further out to sea. During the latter part of the Mesozoic Era (Cretaceous Period) and into the Cenozoic Era (Tertiary Period), sea level was fluctuating due to climatic changes and alternately submerging and exposing the coastal plain. These changes are today reflected in the sediments of that area. They're essentially still sediments since they weren't buried quite deeply enough or long enough to lithify into sedimentary rocks.

Coastal Plain sediments, formed during the Late Cretaceous Period about 90 million years ago, are represented in New York State primarily by the Raritan Formation. "Raritan" was the name of a tribe of Lenape Indians who inhabited Staten Island and adjacent areas of New Jersey to the west. The Raritan River, which enters into the bay of the same name near the southern shore of Staten Island, played an important role in the evolution of our own Hudson River as we'll see in a later chapter. Near the end of the Cretaceous, the Coastal Plain was characterized by a series of marine transgressions and regressions alternately flooding and exposing these low-lying areas leaving behind terrestrial, coastal, and nearshore marine clays, sands, and gravels. Some of the members of the Raritan Formation also contain abundant plant and animal fossils providing us with a window into life at the end of the Mesozoic Era.

The Raritan Formation runs across the southeastern half of Staten Island and then across the Verrazano Narrows into Brooklyn and Queens on the western end of Long Island. While the Raritan has been extensively studied in New Jersey, it's not as well-known on Staten Island and Long Island since it's covered by younger glacial deposits that we'll learn about in the next chapter. It is sporadically exposed wherever excavations are dug, and along bluffs on the northwestern shore of Long Island. Different members of the Raritan have some economic uses; clay layers have been used for pottery and brick making and sand layers have been mined for use in the construction industry. Sand layers within the Raritan also form important drinking water aquifers in the subsurface of Long Island since they store a lot of fresh water.

There aren't many good places to easily examine the Raritan Formation since it underlies a heavily-developed part of the country. Kennedy Park, a city park in Sayerville, New Jersey does contain some exposures of the Raritan Formation although the area has been heavily disturbed by past mining for clay. The clay made excellent fire bricks — bricks resistant to high temperatures and used to line furnaces and kilns. Sand was also mined from this area for use in the construction industry. The area surrounding this park was once a favorite haunt for those collecting amber from the Raritan Formation back when it was less developed. Amber is fossilized tree sap, typically from conifers like pines. It's relatively easy to see how such deposits formed when, even today, balls of tree sap from pitch pines (Pinus rigida) can accumulate in some New Jersey shoreline deposits after large coastal storms. The great thing about preservation in amber is that some of the soft tissues of the insects are preserved allowing for more detailed study (but no intact DNA, alas, as it's a fragile molecule that breaks down relatively quickly despite the central premise of Jurassic Park). Lignite, which is partially carbonized wood on its way to becoming bituminous coal, is also abundant in the area.

These Raritan Formation plant fossils are important because a significant advance occurred in the plant kingdom during this time. Around 100 million years ago, in the Middle Cretaceous Period, we see the advent of flowering plants called angiosperms. The prefix *angio* is Greek for "case" or "container" and sperm means "seed." These are the enclosed-seed plants as opposed to the naked-seed plants, or gymnosperms, that were dominant from the Carboniferous into the Mesozoic Era. When hearing the term "flowering plants," it's typical for people to think of plants with pretty flowers, but this large group of plants encompasses far more. Angiosperms make up about 80% of all plants on Earth today and include small herbaceous plants, grasses, shrubs, and deciduous trees (trees that seasonally lose their leaves). All of these plants, if you look carefully enough, have flower-like reproductive structures. Prior to the latter part of the Cretaceous Period, there were no grasslands on Earth. Dinosaurs tromped through plains of ferns and cycads. Familiar trees like oaks, maples, birch, and willow did not exist and large conifers dominated the landscape. With the appearance of flowering plants, the terrestrial landscape became truly colorful for the first time in Earth history.

In the animal kingdom, there were two major advances during the Mesozoic Era, the advent of birds and mammals. Both of these large groups evolved independently from different lineages of reptiles.

Birds evolved from the bipedal, carnivorous dinosaurs called theropods. While most people think of the imposing T. rex when imagining such dinosaurs, many of them were the size of ostriches or even smaller. There were also many bird-like dinosaurs which evolved light, hollow bones and feathers (feathers are modified reptile scales). Another important characteristic these dinosaurs evolved, that's not directly represented in the fossil record, is warm-bloodedness or endothermy. Endothermy allows an organism to have greater stamina, a useful characteristic when running down prey or running away from a predator.

The Jurassic Period fossil Archaeopteryx, several of which have been found from the famous Solnhofen Limestone in the Bavaria region of southern Germany, is a perfect example of a dinosaur-bird transitional fossil having characteristics of both groups. While having the reptilian features of teeth and a long, bony tail, it also had feathers, wings, and a wishbone, all avian features. By the Cretaceous Period, at the end of the Mesozoic Era, we start to find fossils of true birds, albeit ones still retaining some reptilian characteristics like small teeth in their beaks.

Mammals, on the other hand, did not evolve from dinosaurs at all but from a completely different group of reptiles called the therapsids (don't confuse them with the theropod dinosaurs). Therapsids are sometimes called the mammal-like reptiles because we see in them a clear evolutionary progression from reptiles to mammals. Therapsids first arose in the Permian Period, at the end of the Paleozoic Era, as lizard-like reptiles. They survived the great Permian extinction and gradually evolved increasingly more mammalian characteristics as they moved through the Triassic Period eventually evolving into middle Mesozoic mammals.

Mammal fossils have a number of skeletal features which help us to distinguish them from those of reptiles. The most important is differentiated teeth. If you carefully look into the mouth of a reptile, like a crocodile, you'll see that all of their teeth look essentially the same. If you look into a mirror at your own teeth, you'll see that you have differently-shaped teeth for different functions – incisors for tearing, canines for stabbing (not well-developed in humans, but check out your dog or cat's mouth), premolars and molars for chewing and grinding. Different groups of mammals have different numbers of differentiated teeth. Carnivores have long canines and scissor-like premolars for eating meat, cows and horses have large, flat molars for grinding grass, while rodents have elongated incisors for gnawing.

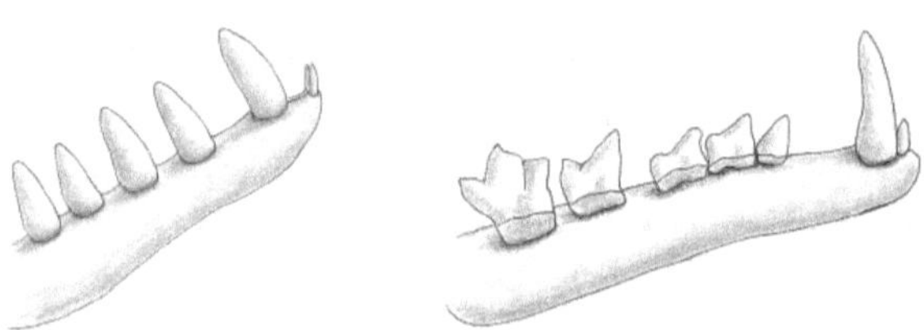

Undifferentiated teeth in a reptile jaw (left) versus differentiated teeth in a mammalian jaw (right)

While there are several skeletal features besides differentiated teeth which allow us to distinguish mammals from reptiles, there are also other soft-tissue features not so easily fossilized. Mammals have fur, rather than reptilian scales, and mammary glands (from which mammals get their name) to produce milk for their young. Mammals, like birds, are also warm blooded. This metabolic trait apparently evolved separately in both mammals and birds from different groups of reptiles. Exactly when endothermy appeared is a bit difficult to figure out although a few tantalizing clues are preserved in fossilized bones leading some paleontologists to believe that some reptile groups were at least partially endothermic as well.

We do see primitive mammals, however, by the latest part of the Triassic Period. These mammals were small, rat-like creatures that lived as unobtrusively as possible in burrows and presumably spent their time trying to avoid being eaten by theropod dinosaurs. The middle of the Mesozoic Era was known as the "Age of Reptiles" as these creatures were the dominant animals on land (dinosaurs), in the seas (marine reptiles), and even in the air (pterosaurs). The modern Cenozoic Era is sometimes called the "Age of Mammals" because of the dominance of mammals in today's world with reptiles relegated to a minor role in most ecosystems. What happened to effect this dramatic change?

As Pangaea rifted apart during the Mesozoic Era, the world started to resemble the one we're familiar with today. A new passive margin formed on the East Coast of North America, the first since the Cambrian Period a few hundred million years earlier. The area that's now the Hudson Valley was above sea level and an area of erosion. No sediments were being deposited. This is frustrating for a geologist since we know that a lot of interesting things were occurring in the history of life, but no fossils recording these events were left here in New York.

We have already discussed the rare dinosaur tracks found in southeastern New York. Dinosaurs were presumably tramping all over New York State during the Mesozoic Era along with the earliest mammals but, like the footprints you may leave today along muddy Hudson Valley hiking trails, these traces were not destined for permanent preservation.

Dinosaurs abruptly ended their almost 180 million year reign on Earth around 65 million years ago. This is why, similar to the mysterious Permian extinction event, there was a boundary placed in the geologic time scale between the Mesozoic and Cenozoic Eras. Since this is also the break between the Cretaceous and Tertiary Periods, it's often called the K-T boundary (K is the abbreviation for Cretaceous since C is used for the Carboniferous). We believe, however, that we have a better handle on what caused this more recent mass extinction event.

In 1980, physicist Luis Alvarez, along with his geologist son Walter and two chemists, published a landmark paper in the journal Science where they speculated on an extraterrestrial cause for the extinction of the dinosaurs. They were studying a sequence of limestones in Gubbio, Italy and noticed a spike in the concentration of rare-Earth elements (REEs) at the K-T boundary. As the name implies, REEs are rare in Earth rocks but a common constituent of space dust that's constantly raining down into the Earth's atmosphere and becoming incorporated into seafloor sediments at a relatively constant rate. Alvarez and his team were hoping that by measuring the concentrations of REEs in the limestones across the K-T boundary, they could use that as a clock to measure the rate of change in certain marine invertebrate fossil groups found within those rocks. Instead what they found was a huge spike in the amount of REEs like iridium and osmium at this boundary.

Luis and Walter Alvarez at the K-T boundary in Gubbio, Italy (1981)

It turns out this anomaly can also be measured in such widely-separated localities as Denmark, New Zealand, and Southern Colorado. In addition to this geochemical anomaly, glassy spherules, shocked quartz, and even soot particles were identified from this boundary layer at locations around the world. We've already discussed the significance of glassy spherules and shocked quartz in Chapter 9; they're typically of meteoritic origin. The rare-Earth elements iridium and osmium are also something that might be expected from a meteoric impact since the elements are far more common in extraterrestrial material than they are on Earth. Soot could have resulted from massive forest fires triggered by the impact event.

When Alvarez and his team published in 1980, they proposed that a large meteor impact left these geochemical signatures and may have resulted in the extinction of the dinosaurs. One thing they didn't have was a meteor crater – a smoking gun for the impact hypothesis. By 1990 however, a crater of the right size and age was widely recognized to exist in the subsurface sediments off the northern coast of the Yucatan Peninsula in Mexico. The crater had actually been discovered years earlier by oil-company drilling, but the information was not widely disseminated at the time since it was proprietary. This impact crater was named Chicxulub (usually pronounced "cheek-shoo-loob" although variations exist) after a small Mayan town in the area (the name means "tail of the devil").

Impact site of Chicxulub on the edge of the Yucatan Peninsula of Mexico

Judging by the size of the impact crater, geologists estimate that the asteroid which struck there 65 million years ago was about 10 kilometers (6 miles) in diameter. An asteroid of this size will strike the Earth, on average, about once every 100 million years but, when it does, it's a catastrophic event. As the asteroid hit the shallow sea covering this area at the end of the Mesozoic Era, it would have barely slowed down before striking the limestone bedrock of the seafloor excavating a massive crater. A huge volume of limestone would have been vaporized, liberating large amounts of carbon dioxide (CO_2) gas into the atmosphere along with massive volumes of dust. The carbon dioxide would have formed carbonic acid (H_2CO_3) in the atmosphere and resulted in acidic rainfall. Contained within that dust were the physical and geochemical signatures of the impact which would eventually settle throughout the world – including the location at Gubbio, Italy.

Much of our speculations about what happened during this K-T impact event are based upon studies done in the 1980s on the possible environmental impacts of an all-out nuclear war with the old Soviet Union. A major nuclear war would eject massive volumes of dust into the atmosphere (the difference being that it would be radioactive dust) resulting in what researchers termed a "nuclear winter" scenario. When dust from nuclear explosions, asteroid impacts, or volcanic eruptions reaches the stratosphere at altitudes above 10 kilometers (6 miles), it's caught up by the strong jet-stream winds and spread around the globe in a relatively short period of time. The dust will remain in the stratosphere for weeks or months until it's eventually flushed out by precipitation. While it's up there, however, it scatters and reflects some of the sunlight coming into the Earth's atmosphere. This will act to reduce global temperatures on the surface. Large amounts of dust will reduce temperatures by significant amounts, hence the "nuclear winter" epithet.

Imagine the Late Cretaceous world. Dinosaurs like T. rex and Triceratops woke up to a day like any other day. Then a flash in the sky followed minutes to hours later by a devastating shock wave. The fortunate dinosaurs were killed instantly. The intense heat of the blast started widespread forest fires which killed many more of the animals in the days to follow. Shortly after the impact, it began to get dark and it stayed dark for weeks. With the darkness came the cold. For many of the dinosaurs, unable to regulate their body temperatures or hibernate, the cold was lethal. The acidic rain, darkness, and cold would have killed off much of the vegetation resulting in widespread starvation.

By the time it all over, the dinosaurs were extinct. A few reptiles survived, the crocodilians and some lizards, but the great marine reptiles and the many species of flying reptiles (pterosaurs) were unable to hold out. Other plant and animal species were unable to survive as well — all told about half of all the species on Earth became extinct and possibly up to 90% of the marine plankton. Mammals, however, survived this extinction event. After the extinction of the dinosaurs, mammals were able to be fruitful and multiplied — soon occupying all of the ecological niches made empty by the extinction of the dinosaurs.

We can only guess what types of mammals were roaming New York State during the Tertiary Period. There are a few localities in North America that have preserved the diversity of mammals during this time. One such place is the White River Formation best exposed in the Badlands of South Dakota. It's impossible to hike here without tripping over teeth and bones weathering out of the soft siltstones that formed in an ancient floodplain environment some 30 million years ago. Pig-like entelodonts, sheep-like oreodonts, and even primitive camels and rhinos grazed the grasslands here during the Tertiary. It's possible that some of these mammals grazed in New York as well, but we'll never know since, sadly, we have no geologic record of this time in our area. No rocks coming "down to us from a foreworld" as John Burroughs so eloquently wrote at the beginning of this chapter. We'll just have to make do with clues we can read elsewhere in the world.

As we move into our more modern world, the continents are very close to their present-day positions. The climate, however, was changing rather drastically. One thing the mammals of New York would have noticed near the end of the Tertiary Period was that it started to get cold. Snow wasn't completely melting in the summers. Food was becoming more difficult to find. Life was getting hard. Soon winter would come – a winter that would last, on and off, for almost two million years.

13
The Ice Age

The more it
SNOWS– tiddely-pom
The more it
GOES– tiddely-pom
The more it
GOES– tiddely-pom
On
Snowing.

A. A. Milne
The House at Pooh Corner

It started snowing a little over two million years ago. And, as in Winnie-the-Pooh's song, the more it snowed, the more it kept on snowing. As the global climate cooled, the deep snows didn't totally melt away during the short summer months and each new winter increased the amount on the ground. As larger areas were permanently covered with snow, more of the Earth's surface reflected away sunlight which exacerbated the cooling. As snow accumulated over tens of thousands of years, it compacted into glacial ice and then spread outward under its own weight, like a puddle of honey spilled onto a kitchen table. From the Laurentian Highlands of Quebec, glaciers of the Laurentide ice sheet began to relentlessly move southward into eastern New York. The Earth was, once again, in the throes of a massive ice age. We've never really left it, as a matter of fact, even though global temperatures are currently increasing. A small remnant of the Laurentide ice sheet still exists on Baffin Island in far northeastern Canada and most geologists believe the glaciers will advance again in the geologically-near future – although anthropogenic (human-caused) climate change is complicating that scenario.

Snow crystallizes from water vapor in the atmosphere and most of that water vapor is ultimately derived from the evaporation of seawater (the oceans, after all, cover 71% of the Earth's surface). As seawater evaporates, and moisture precipitates out of the atmosphere as snow, global sea levels will eventually drop if that snow doesn't melt into liquid water and find its way back into the oceans. During the height of the most-recent ice age, approximately 20,000 years ago, sea level was over 100 meters (over 300 feet) lower than it is today. Large areas of what is now underwater as the continental shelf were then exposed as dry land. Next time you're standing on a beach at the Jersey shore, imagine grasslands stretching tens of kilometers further east and a lumbering mammoth grazing placidly in the distance.

There are basically two types of glaciers. Massive sheets of ice which grind across the landscape are called continental glaciers, and smaller "rivers of ice" which flow from high to low elevations in mountainous areas are alpine or mountain glaciers. Geologists divide these into other subtypes but we won't worry about them here.

Continental glaciers only exist today in Greenland and Antarctica where vast sheets of ice, up to 3 kilometers (almost 10,000 feet) thick, have spread slowly outward over time. At the edge of the land masses, they form ice shelves over the surrounding oceans where they eventually calve off into icebergs. Continental glaciers can only form over land, not water. The vast sheet of ice which covers the Arctic Ocean, for example, is simply sea ice, not a glacier, and doesn't get much more than 3 meters (10 feet) in thickness. One important implication of this is that plate tectonics played an important role in glaciation in the Earth's geologic past. If there were no large areas of land mass at high latitudes (near the poles), continental glaciers would generally not form unless the climate became extremely cold (remember Snowball Earth?).

Mountain glaciers are also found on Earth today in a number of locations. The world's major ranges like the Himalaya, Alps, Andes, and Rockies all contain permanent glaciers on some of the high peaks which flow down to lower elevations where they melt away. Since the Earth's climate has been warming over the past century, these glaciers have been retreating further and further up the mountains each year. Because of global warming, some scientists have even predicted that within a couple of decades we'll no longer have glaciers in Montana's Glacier National Park.

Both types of glaciers were present in New York State at various times during the Pleistocene Epoch. The Pleistocene spanned between 2.6 million and 12,000 years ago although the exact start and end times are subject to revision since different researchers will have their own way to define this epoch of geologic time. During times of extreme cold, there were massive continental ice sheets covering the entire area. At other times, there may have been only mountain glaciers in peaks of the Adirondacks to the north. During warmer periods such as we have today, there were no glaciers found anywhere in the state. Studies of glacial sediments, especially in the Midwestern United States, have shown that there were at least four major pulses of glaciation during the Pleistocene. Each resulted in the southward movement of continental ice sheets and they were separated by warmer periods called interglacials which lasted for tens of thousands of years.

These four advances of glaciation are traditionally called, from oldest to youngest, the Nebraskan, Kansan, Illinoisan, and Wisconsin advances. Most geologists today, however, tend to use the term "Pre-Illinoisan" rather than Nebraskan and Kansan because of the difficulty of uniquely assigning older glacial deposits to either of these two events. In actuality, geologists have recognized that the history of ice advances and retreats during the Pleistocene was far more complicated than this simple picture would lead you to believe. A major difficulty in glacial chronology is that younger glacial events have, in many cases, completely overprinted older ones. This is analogous to the problem geologists have had in attempting to unravel Taconic, Acadian, and Alleghenian structures in the Appalachians.

These four major glacial advances are also recognized in Europe where they've been called the Günz, Mindel, Riss, and Würm respectively. In between the glacial advances were the warmer interglacials. In North America, they're called the Aftonian, Yarmouth, and Sangamon stages from oldest to youngest. In addition to these major events, studies have shown that there were dozens of cycles of warming and cooling within each of these glacial and interglacial episodes of the Pleistocene Epoch. These climatic changes were reflected by fluctuations in the temperature of ocean water throughout this epoch.

How do geologists know what the temperature of the oceans was tens of thousands of years ago? It turns out that ocean temperatures are recorded in the shells of microscopic zooplankton called foraminifera (forams for short). Forams create small calcite ($CaCO_3$) shells, called tests, from calcium ions (Ca^{2+}), water (H_2O), and dissolved carbon dioxide gas (CO_2) in seawater. When forams die, their tests settle down through the water column and are eventually incorporated into the seafloor sediments where they're preserved as microfossils.

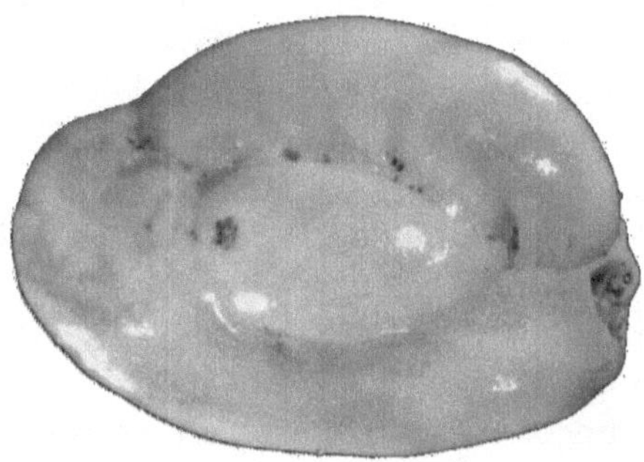

Electron microscope image of foraminifera (Quinqueloculina) from the North Sea. Shell is approximately ½ millimeter in width.

Earlier in the book we discussed isotopes which are atoms of the same element, like carbon, which exist in different forms due to having different numbers of neutrons in their nuclei. Oxygen isotopes also exist in the natural environment. Standard oxygen (99.8%) is oxygen-16 (^{16}O). The mass number of 16 denotes that standard oxygen has a total of 16 protons and neutrons in the nucleus. Since oxygen has an atomic number of 8 on the period table of the elements, it will therefore have 8 protons and 8 neutrons in the nucleus. One naturally-occurring isotope of oxygen (0.2%) is oxygen-18 (^{18}O). This isotope also has 8 protons, as does all oxygen, but requires 10 neutrons to total the mass number of 18. This is termed a heavy isotope of oxygen since, having two additional neutrons in the nucleus, it has more mass. It's also a stable isotope which means it doesn't radioactively decay. The concentration of ^{18}O isotopes in a sample are measured by an instrument called a mass spectrometer and usually reported as a ratio ($^{18}O/^{16}O$) to standard ^{16}O.

It turns out that colder seawater has slightly more ^{18}O in it than warmer seawater. The actual mechanisms resulting in this fractionation of isotopes are complicated, but suffice to say that it takes more thermal energy (higher temperature water) to evaporate the heavier ^{18}O isotopes out of seawater. The bottom line is that foram tests formed in colder seawater have a larger ($^{18}O/^{16}O$) ratio than foram tests formed in warmer seawater. Researchers can thus take foram tests of different ages from deep sea drilling cores, run them through a mass spectrometer, and then apply a formula to the reported ($^{18}O/^{16}O$) ratio to calculate the water temperature in which they lived. Thousands and thousands of samples later, from multiple locations around the world, allow scientists to construct a curve showing how ocean water temperatures have changed through time during the last ice age.

What caused these dramatic changes in climate over the past 2 million years? One of the most important controls on the timing of glaciation is thought to be astronomical cycles which periodically change the amount of insolation (incoming solar radiation) over time. These cycles were first mathematically described in 1920 by Milutin Milanković, a Serbian civil engineer and mathematician, and are thus termed Milankovitch cycles (note the slight variation in spelling). Milankovitch cycles are composed of three different superimposed cycles in the Earth's orbital shape, tilt, and the wobble of its axis.

The first component of the Milankovitch cycles is the Earth's orbital eccentricity. The first law of planetary motion, developed by the German mathematician Johannes Kepler around 1605, states that all planets orbit the Sun in an ellipse, not a perfect circle as everyone had previously believed. The shape of an ellipse is described by its eccentricity – a number between 0 and 1 where an eccentricity of 0 describes a circle and 1 is an open parabola. The eccentricity of the Earth, while it's currently around 0.02, varies from about 0.005 to 0.05 over a scale of around 100,000 years. This change in eccentricity is a result of gravitational interactions with other planets in our solar system, primarily the gas giants Jupiter and Saturn.

Many people are under the misconception that the distance from the Earth to the Sun is what causes seasons. This is easily disproven by the simple fact that the Earth is closest to the Sun (a position called perihelion) in early January while it's furthest away (aphelion) in early July. With the current orbital eccentricity, the change in distance from the Earth to the Sun varies by about 3.5% which isn't enough to have much of an effect on the amount of insolation. At times of higher orbital eccentricity, on the other hand, this variation in distance is much greater and can play a role in climate change by lengthening the seasons.

The second component of the Milankovitch cycles actually is the primary cause for seasons on Earth. The Earth rotates daily on its axis as it travels on its yearly orbit around the Sun. If the axis of the Earth were perfectly perpendicular to the plane of its orbit around the Sun, called the plane of the ecliptic, we would say that it's axial tilt is 0°. In actuality, however, the Earth's axis is tilted by about 23.5°. This axial tilt cycles between 22° and 24.5° over a cycle with a period of a bit over 40,000 years.

This tilt of the Earth's axis causes our seasons. For half the year the Northern Hemisphere is tilted toward the Sun and for the other half it's tilted away from the Sun. The time each year when the Northern

Hemisphere has its maximum tilt toward the Sun is called the summer solstice and generally occurs on June 21. The opposite occurs six months later, around December 21, and is called the winter solstice. The halfway points, when the Northern Hemisphere is neither tilted toward nor away from the Sun, are called the vernal (spring) and autumnal (fall) equinoxes and occur in March and September respectively. The seasons are simply caused by the fact that when the Northern Hemisphere is tilted toward the Sun, the Sun's rays come in more directly and days are longer. Those two factors result in more heating of the Earth's surface at this time. The opposite occurs during the winter months when the Sun's rays are more oblique and days are shorter.

The final component of the Milankovitch cycles has to do with the wobble of the Earth's axis. Like a spinning top slowing down over time, the Earth's axis traces out a cone over a cycle spanning a little over 20,000 years. Currently, the northern axis of the Earth points roughly toward a star we call Polaris or the North Star in the constellation Ursa Minor. This is only a fortuitous coincidence of the time. Around 2,500 BCE, when the Egyptians were building the Great Pyramid of Giza, a dim star called Thuban in the constellation of Draco was the north star. In another 12,000 years, the bright star Vega in the constellation Lyra will serve that function. This wobble of the Earth's axis is also called the precession of the equinoxes since it was historically determined by ancient Greek astronomers observing the position of the spring equinox against the background stars of the zodiacal constellations.

All three components of the Milankovitch cycle can be mathematically summed together to derive a complicated curve of solar insolation which is thought to play an important role in governing the timing of glacial advances and retreats throughout the Pleistocene Epoch. Unfortunately, this is not the complete story since glacial ice movements don't perfectly correlate with these cycles. In addition, throughout the Mesozoic and early Cenozoic Eras, there was no appreciable glaciation even though Milankovitch cycles were occurring just as they do today. What else has acted to trigger glaciation?

Plate tectonics certainly plays an important role in the timing of glacial ice ages. As mentioned earlier, glaciation won't even occur unless there are areas of continental crust at high enough latitudes to accumulate glacial ice. But, in addition to this, plate tectonics also plays an important role in triggering long-term climatic changes. Two important tectonic events which have altered global climate recently, in a geologic sense, are the creation of the Himalayan Mountains and the closure of the Ismuth of Panama.

The collision of India with Asia began around 70 million years ago near the end of the Mesozoic Era and much of the growth of the Himalayan Range occurred during the Tertiary Period of the Cenozoic Era. Some climatologists have speculated that growth of large mountains may result in global cooling since the chemical weathering of silicate minerals removes carbon dioxide, a greenhouse gas, from the atmosphere.

Another important tectonic event occurred just before the Pleistocene ice age when the Ismuth of Panama closed separating the Atlantic from the Pacific Ocean. This closure strengthened the Gulf Stream and resulted in the Atlantic Ocean becoming saltier than the Pacific, mainly due to evaporation of seawater from dry winds off arid North Africa. As this saltier water is brought into the North Atlantic by the Gulf Stream, it becomes dense enough to sink and creates a deep current of water that oceanographers call the oceanic conveyor belt. The deep flow of water in this system, called a thermohaline current, is very important in the regulation of our present-day climate. Not only do these currents circulate heat energy through the ocean system, they also remove carbon dioxide gas from the atmosphere and sequester it in deep ocean waters for hundreds of years. Prior to the closure of the Ismuth of Panama, however, this water was not as saline, making it less dense, and therefore able to move into the surface waters of the Arctic Ocean before sinking. This movement of warm water into the Arctic apparently helped to keep that area ice-free prior to the Pleistocene.

Related to plate tectonic activity is volcanism. Volcanic eruptions during the Pleistocene Epoch may have also played a role in triggering episodes of shorter term cooling or longer term warming. Ash-rich volcanic eruptions cause several years of cooling due to ash getting into the stratosphere and being spread around the Earth by jet-stream winds. This ash will act to reflect some of the solar insolation, lowering global temperatures, until the ash is eventually removed by rain or snow since it acts as condensation nuclei for precipitation. On a longer time scale, erupting volcanoes emit the greenhouse gas, carbon dioxide (CO_2). If large-enough quantities are introduced into the atmosphere, it can increase global temperatures for longer periods of time and these events may be implicated in some of the warming periods during the last ice age.

There were a number of significant volcanic eruptions during the Pleistocene Epoch that certainly affected global climate at the time. The largest such eruption took place around 70,000 years ago at Lake Toba in the northern part of the island of Sumatra in Indonesia. Indonesia is famous for its volcanoes, such as Krakatoa, because the seafloor of the Indian Ocean is currently subducting beneath this island nation. As the oceanic crust subducts, it melts and provides magma for the growth of these volcanoes. The subduction also causes large earthquakes – a massive earthquake along this subduction zone resulted in the December 2004 Indian Ocean tsunami which killed over 200,000 people.

The eruption of Toba was one of the largest volcanic events in Earth history emitting some 2,800 cubic kilometers (670 cubic miles) of material. For comparison, Mount St. Helens in 1980 erupted about 1.2 cubic kilometers (0.3 cubic miles). Toba's eruption would have reduced average global temperatures by several degrees for a number of years which may have been enough to trigger an advance of glacial ice. One of the interesting, but controversial, speculations about the eruption of Toba is that there's evidence from studies of mitochondrial DNA that Homo sapiens passed through an evolutionary bottleneck about this time. In other words, it's quite possible that our ancestors on Earth almost became extinct from the climatic changes wrought by this massive eruption.

Another factor, still not well understood, is the possibility that energy output from the Sun may vary over long time scales. One intriguing observation of this has to do with the 11-year sunspot cycle. Every 11 years, the number of sunspots visible on the surface of the Sun varies from a maximum, to a minimum, and then back to a maximum again. This is due to periodic changes in the circulation of ionized gases within the Sun since sunspots result from variations in the Sun's magnetic field. During periods of high sunspot activity, the Sun is more active and this increases the amount of charged particles striking the outer atmosphere of the Earth. The 11-year cycle, however, is thought to be short enough such that the Earth's climate doesn't appreciably warm or cool in response to these changes in the solar wind.

Galileo was the first scientist to systematically observe sunspots in 1609 with one of the earliest telescopes. Between around 1645 and 1715, however, a funny thing happened to sunspots – they essentially disappeared. This period of time is called the Maunder minimum after a solar astronomer who studied historical records of sunspots in the late 1800s. It turns out that the Maunder minimum also corresponds to one of the coldest parts of a historical period known as the Little Ice Age. It's important to mention, however, that there is still a lot of debate among climatologists regarding any possible causal relationship between these two events. The bottom line, however, is that we've only studied the Sun for a miniscule percentage of its several-billion-year history and variations in solar output may well have played a very important role in Earth's long term climatic changes.

Prior to the 1800s, glacial deposits were observed in various places throughout Europe and eastern North America but were traditionally ascribed to material deposited by Noah's flood. The geologic term "glacial drift" is a holdover from that earlier time when it was thought that such sediments were drifted in by floodwaters. In 1840, the young Swiss zoologist and geologist Louis Agassiz published a groundbreaking book (*Etudes sur les Glaciers*) in which he proposed the radical idea that Europe was once covered by sheets of glacial ice just as Greenland is today. Agassiz based this claim on studies of active glaciers and glacial deposits in the Alps and seeing these types of deposits elsewhere in Europe. Many of the terms for glacial features today have French names because of this early study of glaciation by French-speaking scientists in the Alps. While his ideas received criticism for a time, he challenged doubters to come and study the sedimentary deposits left by glaciers in the Alps for themselves and gradually won over his opponents. This is a beautiful demonstration of how new ideas are ideally accepted in science, not by authority or passion, but by convincing other scientists with evidence that they too can study and use to reach the same conclusion.

Agassiz moved to the United States in 1846 and was appointed professor at Harvard University. He traveled extensively throughout North America for the next few decades and continued to gather evidence of glaciation and win over adherents to his idea of a global ice age. Agassiz even traveled in the Hudson Valley visiting places like New York City, West Point, and Albany. His confidence in his theory would certainly have been reinforced by the glacial features present throughout our area. The idea that glacial sediments were deposits from a global flood, and the very idea of such a deluge, was finally abandoned in geology in large part due to the research of Agassiz.

Agassiz was a brilliant man, known for his studies in zoology, primarily the study of fish (ichthyology), as well as for being known as the "Father of Glaciology." By the mid-1800s, he was world-famous for his contributions to science. Unfortunately, he's also remembered for being a staunch opponent of Darwinism. This was primarily due to him being a proponent of what's been called "scientific racism" and his belief that blacks and whites were created separately by God (polygenesis) – a belief, of course, that was used to justify the unequal treatment of blacks. Agassiz would have been horrified by the modern idea that we're all evolved from East African hominids. This just goes to show that someone can be truly brilliant in some areas of life and a complete fool in others.

Once the idea of a glacial ice age became established, formerly mysterious features in the landscape were now blindingly obvious. As glaciers advance, they act like bulldozers – Agassiz called them "God's

Great Plough." They roll through the landscape, scouring the sediment and bedrock, and picking up material and entraining it within the ice. When they retreat, however, glaciers act more like dump trucks. There's no physical movement of the glacial ice, it simply melts and flows away. As the glacier melts, all of the material that was carried along by the ice is simply dumped. Some material, of course, is also removed by the significant amounts of water generated by a melting glacier.

There are a number of erosional features glaciers leave behind as they ride across the landscape. Glacial polishing is common and refers to a smoothing of the bedrock from the movement of glacial ice over time. The polishing is not from the ice itself. Go outside and start rubbing an ice cube on a rock if you doubt this. It's from sediment embedded within the ice. This sediment acts like sandpaper, eroding and smoothing as the glacier relentlessly slides over the bedrock year after year.

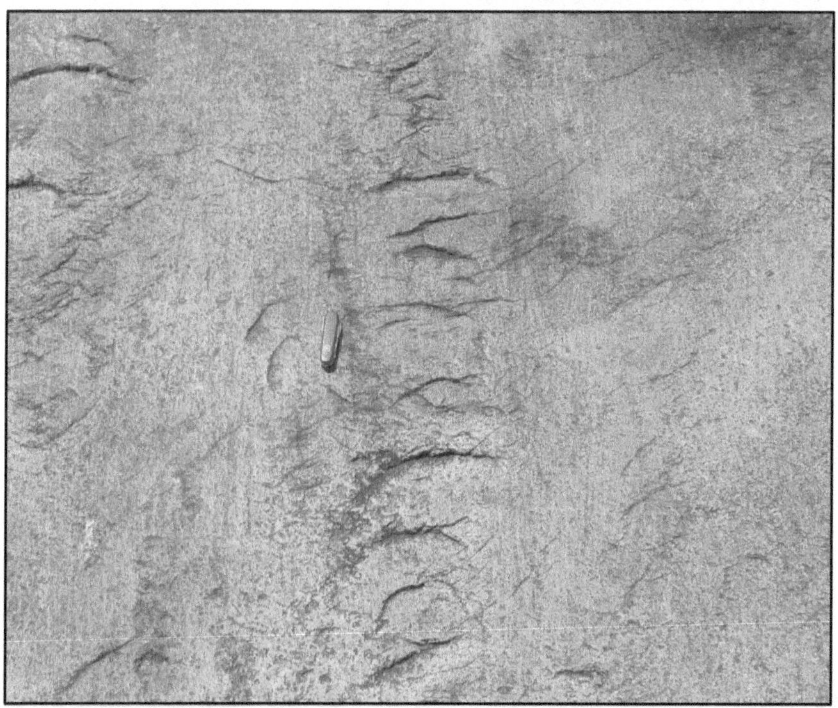

Chatter marks and striations in the Shawangunk Conglomerate at Lake Minnewaska State Park. Swiss army knife for scale.

Glacial striations, which are scratches and grooves carved into the bedrock, are often associated with glacial polishing as well. These gouges are caused by larger fragments of rock embedded in the bottom of the glacier. These features are useful to geologists because they parallel the ancient movement of glacial ice. By measuring and mapping glacial striations over a large area, they can work out the general direction of regional ice advance. Sometimes, rocks embedded within the bottom of the glacier will not slide smoothly along the bedrock carving a striation, but will instead skip along the surface forming crescentic gouges called chatter marks.

Glaciers riding over isolated outcrops of rock sometimes sculpt them into teardrop shapes called roche moutonnées which is French for "fleecy rock" due to their sheepback appearance. Roche moutonnées typically have a smoother, gently-sloping face on the side the glacier came from and a steeper, more ragged face on the lee side where the glacier plucked away material as it rode over the rock. These outcrops typically also show glacial polishing and striations and also give the direction of ice advance by their shape. Roche moutonnées are common throughout the Hudson Valley region.

One of the best places to see glacial polishing, striations, and chatter marks is at Lake Minnewaska State Park in the Shawangunks. The bedrock exposed all around the lake, and actually all over the Shawangunk ridge, is full of these glacial features. The lake itself would not even be there without the area's history of glaciation. There are five beautiful lakes along the crest of the Shawangunks. From north to south, they're Lakes Mohonk, Minnewaska, Awosting, Haseco (which also has the unfortunate name of Mud Pond), and Maratanza. They're sometimes called sky lakes because of their position at the top of the ridge and have limited catchment basins. The lakes exist because glaciers moved across this ridge, as witnessed by the striations and glacial polishing in the bedrock, and carved out weak areas in the rock forming the lake basins.

In a similar manner, but on a larger scale, glaciers also carved out the Great Lakes and New York's Finger Lakes. In all cases, the glaciers took advantage of the fact that the bedrock was weaker along preexisting joints and faults and preferentially easier to erode. This glacial carving also produced a north-south fabric to the landscape of the Hudson Valley. Standing on overlooks in the Hudson Highlands, Shawangunks, or Catskills will reveal an elongate "grain" to the Valley parallel to the Hudson River. The great east-facing cliffs of these ridges are also a direct result of the massive glaciers grinding down the valley.

If you can't make it to the Shawangunks to view glacial striations and polishing, they're common in the Catskills and Hudson Highlands as well. They can also be found throughout Central Park in the heart of Manhattan. One famous landmark showing numerous glacial features is Umpire's Rock on the south side of the Heckscher Ballfields roughly lining up with 63rd Street and 7th Avenue. Umpire's Rock is a large outcrop of the Manhattan Formation having a roche moutonnée shape and a number of glacially-carved grooves and troughs. Walking around the Park with open eyes will certainly allow you to spot numerous other outcrops with glacial striations and polishing as well.

Umpire's Rock in Central Park showing glacial grooves in the Manhattan Schist.

In addition to these erosional features, the glaciers of the last ice age left many depositional features within the Hudson Valley. One of the most important glacial deposits is till, which refers to unsorted sediments deposited by a glacier. These sediments range in size and include clays, silts, sands, pebbles, cobbles, and even large boulders. It simply represents whatever material was picked up by the glacier and deposited in place when the glacial ice melted away. The entire Hudson Valley is blanketed by a layer of till.

One of the interesting features sometimes seen in glacial till is the presence of out-of-place rocks. For example, in my back yard, in Ulster County I've found cobbles of metamorphic rock. There is no exposed metamorphic rock anywhere in Ulster County. What I'm holding in my hand when I pick up such cobbles are pieces of the Adirondacks transported south by the glaciers and dumped in the area by the melting ice sheet several thousand years ago. These out-of-place rocks are called erratics. While some, like the metamorphic cobbles in my backyard, represent a different lithology from the local sedimentary bedrock, others may be a bit more difficult to identify because they haven't been transported such long distances. The easiest to spot are the large boulders sitting anomalously in the fields and forests of the Hudson Valley – rocks that are clearly out of place in their present positions.

Glacial erratics may also show faceting or striations from being carried along in the base of the ice sheet and eroded as they were scraped against the underlying bedrock. Unusual or unique erratics may also give clues as to the direction from which the glaciers advanced. Sometimes they provide clues, but no answers – several diamonds from an unknown source to the north, one as large as 5 carats, have been found in glacial till in the cornfields of Indiana. Unfortunately, no one has been able to trace them to their source which is presumably a buried kimberlite diamond pipe somewhere to the north.

Patterson's Pellet, a large glacial erratic, sitting atop a cliff along the Millbrook Mountain Carraigeway at Lake Minnewaska State Park

Erratics are common throughout the Hudson Valley. When hiking in the Taconics, Catskills, Shawangunks, or Hudson Highlands, it's not uncommon to see large boulders perched atop cliffs and mountain peaks. They're often the same rock type as the local bedrock showing they haven't been transported very far, but are clearly out of place. One well-known example is Indian Rock, an impressively large boulder of Shawangunk Conglomerate at Sam's Point Nature Preserve near Ellenville. Another is Patterson's Pellet in Lake Minnewaska State Park. Large glacial erratics, some the size of small cabins, are commonly seen scattered across the top of this ridge and usually associated with local glacial polishing and striations on the surrounding bedrock.

At the nose of glaciers, where the advance of ice is balanced by yearly melting, glaciers will typically deposit a ridge of till called a moraine. The word comes from a regional French term for ridges of sediment found in valleys throughout the Alps. Since they represent the front end of the glacier, which will often remain essentially in one place for long periods of time, these types of moraines are called end moraines. As glaciers retreat when the climate eventually warms, it's not uncommon to find many parallel end moraines recording the recession of the glacier over time. This leads to two terms for end moraines – the terminal moraine which records the maximum advance of the glacier and multiple recessional moraines which record the glacier's retreat.

Moraines can be relatively small ridges, a few meters high, such as those deposited by mountain glaciers in Alpine valleys and studied by Louis Agassiz. They can also be massive deposits such as those which compose New York's Long Island. Long Island is composed of two end moraines, prominent east-west ridges generally found on either side of the Long Island Expressway, representing separate glacial ice positions.

The southern ridge is the Ronkonkoma moraine. This is generally taken to represent the terminal moraine of the latest Wisconsin glaciation although some geologists have argued that there's evidence for a moraine existing further south on the continental shelf. The Ronkonkoma moraine extends out to Montauk and can then be traced discontinuously on the continental shelf to Martha's Vineyard and Nantucket Island.

The northern ridge of Long Island is the Harbor Hill moraine. This represents a younger recessional moraine formed as the glacier began its retreat when the weather started to warm. The Harbor Hill moraine can be traced along the northern shore of Long Island and from there into discontinuous, eroded ridges in southern Connecticut and Rhode Island to eventually appear as the backbone of Cape Cod in Massachusetts.

The exact chronology of deglaciation in the Hudson Valley is somewhat difficult to work out due to its complexity. At a broad scale, global temperatures rose and the glaciers eventually melted away ending the ice age. At a fine scale, global temperatures fluctuated over several thousands of years resulting in melting of the glacial ice during warmer periods, and readvances of glacial ice during colder periods. Over time, the retreats outnumbered the readvances, and the great continental glaciers disappeared from North America except for places like Baffin Island and Greenland.

Recessional moraines exist in a number of locations up through the Hudson Valley recording the latest deglaciation as the great ice sheets retreated northward while the climate warmed. About 21,000 years ago, the glacial front was situated at Long Island. By 16,000 years ago, it was in Ulster County about half-way up the Hudson River Valley. By 12,000 years ago, the ice was north of Albany. As you can see, it took almost 10,000 years for the Hudson Valley to become deglaciated and it wasn't a smooth process. The climate fluctuated during this time such that several standstills, and even readvances, of glacial ice occurred for short periods of time.

It's difficult to recognize some of these moraines today because they're significantly smaller than the massive pile of sediments that form Long Island and they've been greatly modified by subsequent erosion and human activity. Significant mid-Hudson Valley moraine deposits are located at the northern edge of the Hudson Highlands (Shenandoah moraine), Wallkill, Poughkeepsie, Hyde Park, Pine Plains, Rosendale, and Red Hook. The age of these moraines obviously decreases as we move from south to north.

Another glacial feature formed of till are elongate, asymmetric hills called drumlins, one of the few glacial terms derived from Irish instead of French. The origin of drumlins is still not well understood but they're basically thought to have formed when glaciers rode over a preexisting layer of till sculpting it into hills with their long axis aligned to the glacial advance direction. The asymmetric shape of the hills also records the direction of ice advance with the steeper face of the hill facing the direction the glacier came from and tapering off more gradually in the direction the ice advanced (this is the opposite sense of roche moutonnés).

Drumlins typically occur in clusters called drumlin fields and one of the largest is located in western New York between Syracuse and Rochester where thousands have been identified. Drumlins often form isolated, prominent hills and thus become significant geographic locations. Two examples are Bunker Hill in Boston, important during the Revolutionary War, and Hill Cumorah near Palmyra in western New York, where Joseph Smith purportedly found the golden tablets used to write the Book of Mormon. Drumlins are found in a number of locations throughout the Hudson Valley and can often be located by looking for the locations of sand and gravel pits on the sides of elongate hills. Virtually all sand and gravel mines in New York State are associated with some sort of glacially-deposited feature.

Meltwater flowing from glaciers can also form depositional features but the difference is that they're composed of stratified sediments rather than unsorted till. Two such features are kames and eskers. Kames are conical hills of layered sediments which form when glaciers are retreating and the surface of the glacier becomes pitted with depressions as the ice is melting. Meltwater carries sediments into these depressions which are then left behind as hills when the glaciers completely melt. Eskers are sinuous ridges of layered sediments which form when flowing streams develop beneath melting glaciers. While kames are relatively common throughout the Hudson Valley, including one less than a mile from where I teach in the town of Stone Ridge, eskers are less common and generally not well-developed in our area. Good examples of these features, however, are found in western New York.

As glaciers melt away, huge volumes of meltwater are generated. This meltwater typically forms a distinctive topography at the front of the glacier called an outwash plain. When till is deposited, features like moraines, kames, and eskers form, and blocks of ice left behind in the sediment melt away to form small ponds called kettles. So-called 'kame and kettle topography' is typical of former outwash plains and kettle ponds are found throughout the lower elevations of the Hudson Valley.

Sometimes, meltwater becomes dammed by old moraines left further down the valley by the receding glacier. This forms postglacial lakes and a number of large lakes formed in the Hudson Valley over the thousands of years that it took for the great glacial ice sheet to retreat to the north. Glacial lakes first formed in the lower Hudson Valley, dammed by the terminal moraine across the Verrazano Narrows between Staten Island and Brooklyn. Later, as the glaciers moved north of Albany, a large postglacial lake existed in the Hudson and Rondout Valleys dammed by moraines just north of the Hudson Highlands. The largest of these, glacial Lake Albany, may have existed for a couple of thousand years and was tens of meters (over 100 feet) deep.

If you can imagine standing on the east-facing cliffs of the Catskills just after the glaciers left our area, the Hudson Valley would lie below you as a vast expanse of still water – a deep, cold, fresh-water lake formed by the melting of glacial ice. Full of rock flour, the ground-up fine sediments eroded by the glacier as it polished the underlying bedrock, the lake would have glinted a deep aquamarine color in the bright afternoon sunshine. The lake would have deposited clay sediments on its bottom – darker, more organic-rich clays in the summer when the lake was open, and lighter, less organic-rich sediments over the long winter when the lake was frozen. These yearly couplets of clay formed alternating light and dark laminations called varves that geologists today can use to count the number of years a postglacial lake existed in an area.

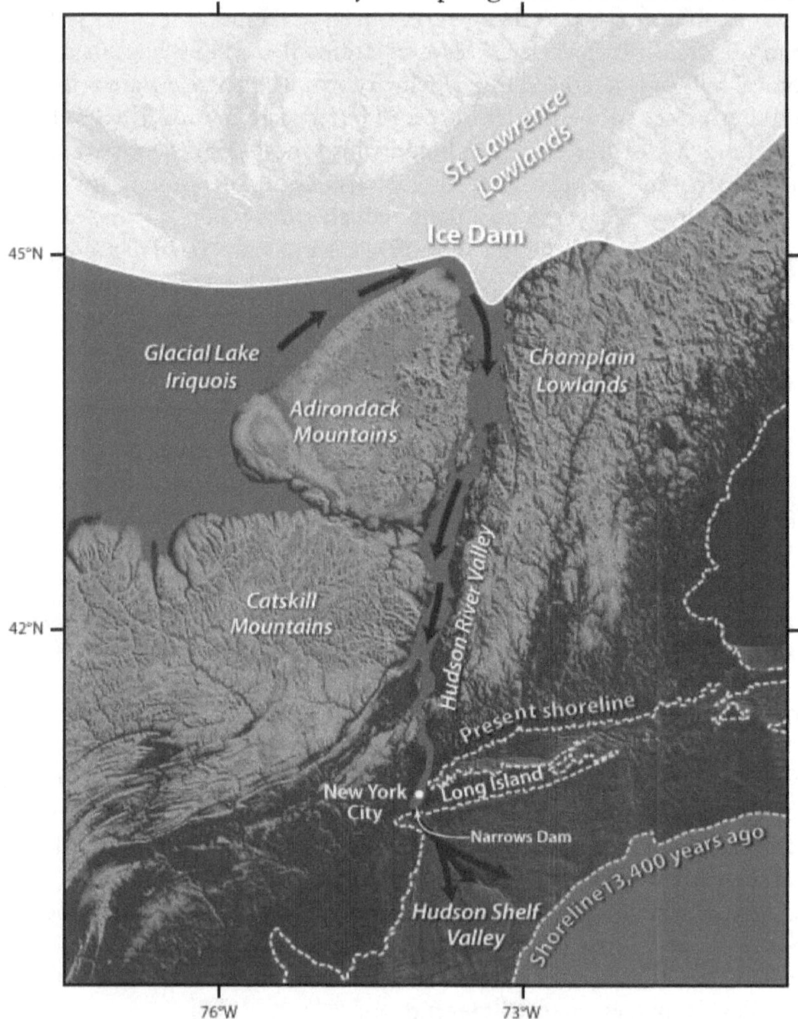

The draining of post-glacial Lake Iriquois through the Champlain Sea and Lake Albany down the Hudson Valley and out on to the exposed continental shelf forming the Hudson Canyon.

These lacustrine (lake-deposited) clays also played an important role in the economic history of the Hudson Valley. Many of these thick clay deposits were suitable for making bricks and supported a thriving industry along the Hudson River from the late 1700s into the early 1900s. Not surprisingly, this was coincident with the building boom in northeastern cities. Their location on the Hudson made shipping easy and, at the height of production in the late 1800s, there were well over 100 different brick-making factories lining the length of the river, many of them clustered around Haverstraw Bay, just south of the Hudson Highlands.

Workers cutting glacial lake clay for brick making in East Kingston around 1900

Throughout the Hudson Valley are deltas which formed when streams flowed down from higher elevations into Lake Albany. These are easily recognized as flat-topped areas of sandy sediments with distinctive internal layering. The elevations of the ancient lake surface can be ascertained by mapping the elevations of the tops of these deltas throughout the region.

Glacial Lake Albany eventually drained to the south, due in part to the uplift of the crust associated with the retreat of the glacial ice. During the height of glaciation, a 1.5 kilometer (5,000 foot) thick sheet of glacial ice covered the Hudson Valley. A massive ice sheet such as this exerts a tremendous weight on the crust and physically depresses it. As the glaciers melt, the crust rises back up, a process called isostatic rebound, as it reequilibrates. Studies have shown that there was over 50 meters (over 150 feet) of rebound in the lower Hudson Valley following deglaciation of the area. Isostatic rebound may have helped glacial Lake Albany erode through the southern moraines blocking its southern edge, and then drain across the outwash plains on the present-day continental shelf and eventually to the sea.

An even larger postglacial lake existed to the north where Lake Ontario is today. Glacial Lake Iroquois was three times the size of the present Great Lake and drained down the Hudson Valley a bit over 13,000 years ago when the waters broke through an ice dam to the north. As the water level dropped by 120 m (400 ft) in this massive lake, torrents of highly-erosive waters would have flowed through our area essentially forming the modern landscape. This huge influx of fresh water into the North Atlantic was thought to have shut down the Gulf Stream current for a time and may have initiated a climatic change called the Younger Dryas Cold Period.

Following deglaciation, flora and fauna quickly recolonized the area of bare rocks and sediments left behind. First lichens, then tundra, and finally conifer forests developed as the climate began to warm. It's almost hard to believe, but just a few thousand years ago, mammoths and mastodons (also spelled mastodonts) were wandering around New York State. While people sometimes use the terms mammoth and mastodon synonymously, they were distinct animals. Mastodons split off from the main line of elephant evolution over 20 million years ago while mammoths arose more recently, less than 5 million years ago. Both lineages essentially died out at the end of the Pleistocene ice age although there's evidence that small groups of mammoths may have existed as recently as 2,000 BCE in isolated northern areas.

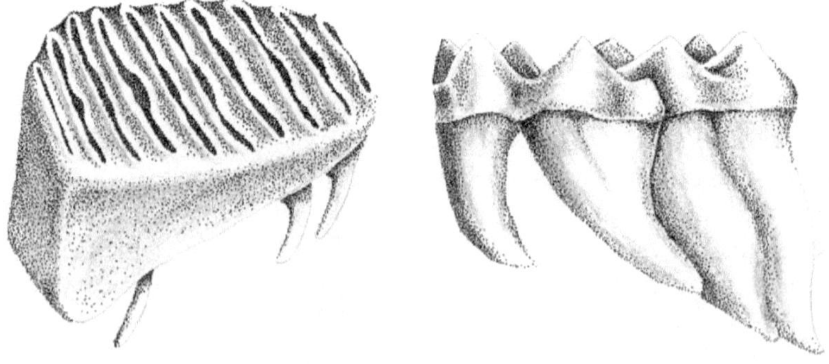

Mammoth (left) versus mastodon (right) teeth

While mastodons are generally shorter and stockier, and mammoths tend to have higher, domed skulls, the most notable difference between mammoths and mastodons is seen in their teeth. Mammoths were grazers, eating grass like modern horses, and had large, flat molars with sub-parallel ridges. Mastodons, on the other hand, were browsers like modern elephants, eating mostly leaves, and had molars more like ours with protruding cusps. The name mastodon literally means "breast tooth" after these bumps on their teeth.

One of the earliest reports of mastodons (*Mammut americanum*) in North America was from teeth and bones discovered around 1705 near Claverack, a small town to the east of the city of Hudson. The Puritan minister Cotton Mather, infamous for his role in the Salem witch trials, wrote a letter to the Royal Society of London in 1712 describing these teeth and bones as the remains from a race of antediluvian giants – presumably the Nephilim of Genesis 6:4. It was not uncommon at the time for Pleistocene mammal remains to be mistaken for those of giant humans or other mythological beasts. The legends of cyclopes, like the one-eyed Polyphemus from Homer's Odyssey, are thought by some to have originated from Pleistocene pygmy elephant remains found on the island of Crete. The skulls of these elephants had a large nasal cavity in the front, where the trunk attached, which might resemble a single eye socket to the casual observer.

Cohoes mastodon mounted in the New York State Museum in Albany

Since the "Clavarack Giant" find, over 100 mastodon sites have been discovered in the Hudson Valley although few have complete (or nearly-complete) skeletons. One famous specimen is the Cohoes mastodon which can be seen in a very nice display at the State Museum in Albany. This skeleton was discovered in 1866 during the excavation for a mill at Cohoes Falls on the Mohawk River. Cohoes is located close to where the Mohawk flows in the Hudson River just north of Albany and the bones, covered with sediment, were discovered in a pothole in the bed of the river. Studies have shown that this American mastodon was an immature male, with bones and teeth showing it to be in ill health, and it died of unknown causes a little over 11,000 years ago.

Another famous Hudson Valley find was the Hyde Park mastodon found in a suburban backyard just north of the city of Poughkeepsie. When the Lozier family decided to enlarge a pond in 1999, the backhoe pulled up a large leg bone. The Paleontological Research Institute (PRI) professionally excavated the find in 2000 and recovered 95% of the bones including the important skull and tusks. It was an adult male, 30-40 years old at death, that stood about 3 meters (10 feet) tall at the shoulder and weighed over 4,500 kilograms (over 10,000 pounds). Radiocarbon dating of the mastodon's bones established that it was walking around the Hudson Valley 11,500 years ago.

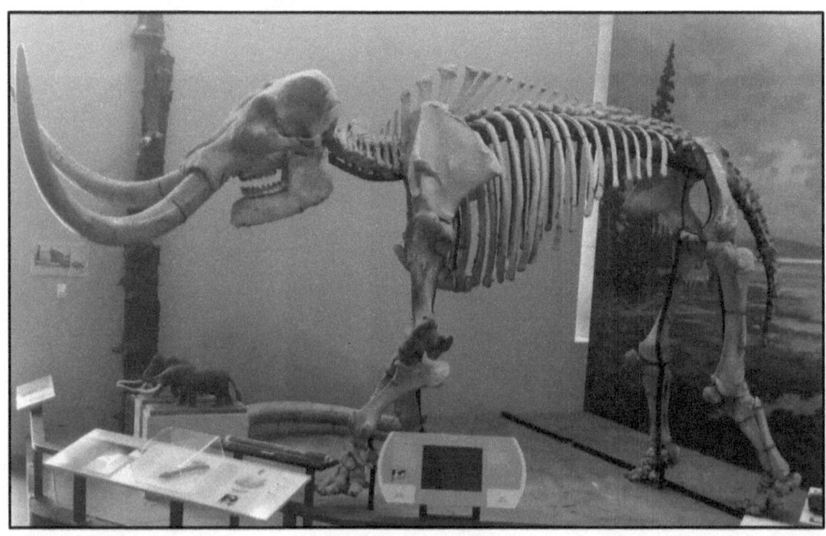

Hyde Park mastodon at the Museum of the Earth in Ithaca

In addition to the excellent mastodon skeleton, the muck in which it was entombed preserved freshwater mollusk and crustacean fossils, insect parts, plant fragments, and pollen; all of which allow paleontologists to reconstruct the flora and climate of the area during this part of the Pleistocene deglaciation of the Hudson Valley. One of the more spectacular fossils, besides the mastodon, was a several meter long spruce log that looked as fresh as if it had been buried yesterday. I had the privilege of visiting the excavation in 2000 and was able to view some of the bones (including the skull) and collect some of the muck from the excavation for my geology lab. Today, the mastodon is on display at the PRI's Museum of the Earth in Ithaca, New York.

A reasonable question to ask at this point might be how we know the age of these mastodons. We've already discussed radioactive isotopes and the radiometric dating of minerals in rocks, but how can we date the remains of once-living organisms? The answer to this question lies in the fact that we're all radioactive – every living thing on Earth.

The atmosphere of the Earth is under a constant bombardment from charged particles called cosmic rays (the term "rays" is a misnomer, they're particles). These particles can originate from vast distances away from us during energetic processes such as supernovae – the explosion of stars as they reach the end of their life span. As these charged particles enter our atmosphere, they interact with nitrogen atoms (remember our atmosphere is 78% N_2), and convert some of them to carbon-14 (^{14}C), the radioactive isotope of carbon.

In the atmosphere, the new radioactive carbon can bond with oxygen molecules (O_2) to form carbon dioxide (CO_2). Atmospheric carbon dioxide is absorbed by plants through photosynthesis, the carbon is used to build plant tissues, and the plants are now radioactive. Animals eating the plants, likewise absorbing carbon atoms into their tissues, also become radioactive. All of us have about one atom of radioactive ^{14}C in our bodies for every trillion atoms of normal ^{12}C. When an organism dies, it no longer absorbs new radiocarbon, and the radioactive decay clock starts counting down.

Radiocarbon, as mentioned in an earlier chapter, decays away with a half-life of 5,730 years. Radioactive dating only works well back to about ten half-lives, more than that and there's little of the original material left to detect. This means that ^{14}C dating is only appropriate for organic material less than about 60,000 years old. Another complication with radiocarbon dating is the fact that the influx of cosmic rays over the past few tens of thousands of years, and thus the production rate of ^{14}C in the Earth's atmosphere, hasn't been constant. Radiocarbon dates thus need to have small correction factors applied to them and researchers have been able to do this by calibrating ^{14}C dates with objects for which they can determine independent ages. One of the most important cross-checks on radiocarbon dates are from studies of tree rings (dendrochronology) but ice cores, coral growth rings, and seasonally-layered (varved) clays have been used as well.

Despite its limitations, however, radiocarbon dating is perfect for dating organic material from more recent parts of the Pleistocene Epoch (as well as being immensely useful for archaeologists). With mastodon sites, radiocarbon dating can be used on the bones as well as associated plant and animal material (the spruce log with the Hyde Park mastodon, for example). Radiocarbon dating can also be used on fragments of wood found in glacial moraines or post-glacial lake clays in order to constrain their ages.

Mammoth remains, mostly consisting of scattered bones and teeth, are rarer in New York than are mastodons. Several wooly mammoth (*Mammuthus primigenius*) remains have been found in western New York but I was unable to come across any confirmed finds in the Hudson Valley. In 1961, however, several bones identified to be from a Jefferson's mammoth (*Mammuthus jeffersonii*) were discovered in a peat bog in Westchester County during road construction on Route 134 near where it crosses the Taconic Parkway.

Giant beavers (*Castoroides ohioensis*) were also living in the Hudson Valley during the Pleistocene Epoch. They were the size of black bears growing up to 2.5 meters (over 8 feet) in length with 15 cm (6 inch) long incisors. Some Native American tribes even have legends of these animals that their ancestors certainly saw when they migrated into North America. Peccaries (*Platygonus compressus*) and massive short-faced bears (*Arctodus simus*), as well as more familiar animals like wolves, elk, caribou, and moose also frequented New York during the last ice age.

Many of these large ice-age mammals, called the Pleistocene mega-fauna by paleontologists, abruptly disappeared around 10,000 years ago. The climate was warming but these animals had weathered similar warming periods during earlier interglacials. Many researchers don't consider it to be a coincidence that the disappearance of these animals roughly coincided with the arrival of Paleo-Indians into these formerly glaciated areas. While there is no direct evidence of Native Americans hunting mastodons here in the Hudson Valley, there are large numbers of mastodon butchering sites elsewhere.

With the arrival of people in the Hudson Valley, we now leave the realm of geologic time and move into the period of human history.

14

Modern Times

When we try to pick out anything by itself, we find it hitched to everything else in the Universe.

John Muir
My First Summer in the Sierra

When we talk about the Hudson Valley, the obvious connecting thread is the river itself. *Muhheakantuck*, Lenape for "the river that flows both ways," has been the heart and soul of the region since the time of its first inhabitants. Even though many maps show the Hudson River beginning at Henderson Lake in the Adirondacks, the furthest headwaters actually flow out of a small pond on the south side of Mount Marcy. Discovered in 1872 by Verplanck Colvin during a major survey of the Adirondacks region, the lake was poetically described as a "tear of the clouds" – a name by which it's still known today. From these humble beginnings, water tumbles down Feldspar Creek into the Opalescent River and then merges with the Hudson River near the old mining town of Tahawus in Essex County. Some 500 kilometers (300 miles) later, these small streams have merged with many others to form the mighty Hudson River flowing southward into the Atlantic Ocean.

The Hudson River Valley is clearly a product of its long geologic history. As we've discussed in previous chapters, the present-day course of the Hudson River was influenced by bedrock geology and carved out by the continental glaciers of the Pleistocene Ice Age. The Hudson today generally divides sedimentary rocks on the western side of the river from metamorphic rocks on the eastern side. Where softer sedimentary rocks are present, the Hudson is a wider, shallower river. The swath of metamorphic rocks forming the Hudson Highlands narrows and deepens the river as it flows through this belt of harder, more resistant rocks. The river itself, however, is a geologically recent feature.

Hudson River Facts	
Length of the River	507 km (315 mi)
Drainage Basin	36,000 km^2 (14,000 mi^2)
Discharge into NY Bay	620 m^3/s (21,900 ft^3/s)
Deepest Point (World's End)	62 m (202 ft)
Widest Point (Haverstraw Bay)	5.5 km (3.4 mi)

What is the origin of the Hudson? While the modern river we know and love only dates back a few thousand years, its precursors reach back prior to the Pleistocene Ice Age. It's difficult to reconstruct ancient drainage systems in areas that were formerly glaciated as was New York. Like in a palimpsest, the ancient parchment manuscripts which were scraped clear of older writings and then reused, it's sometimes possible to recover information that's later been overwritten. In the case of the Hudson River, these erasures were performed by the multiple advances and retreats of massive glaciers significantly rearranging the landscape over the past two million years. If you look carefully, however, you'll find these older rivers left behind a few scattered clues as to where they once flowed.

It's thought that prior to the Pleistocene glaciation, during the Tertiary Period, the Hudson River followed a slightly different course along its southern reaches. After downcutting through the ridges of the Hudson Highlands, following the erodible traces of ancient faults, the Hudson flowed southward to the area just below the modern Tappan Zee Bridge. At Piermont, near Tallman Mountain State Park, the river apparently took a turn to the west where it cut through the resistant rocks of the Palisades Sill at a place now called the Sparkill Gap. Look for the gap to the south the next time you drive westward across the Tappan Zee Bridge. Once again, the river preferentially cut through this area due to the presence of preexisting faults.

Once through the Sparkill Gap, the proto-Hudson had a clear path southward through the lowlands of the Newark Rift. A number of ancient water gaps cut through the basalt ridges of the Watchung Mountains of northern New Jersey and attest to the ancient course of the river in that area. The river then flowed back toward the east, into what's now Raritan Bay south of Staten Island, and from there out into the Atlantic Ocean. Keep in mind that the position of the Atlantic shoreline was almost continually changing, albeit slowly on a human time scale, due to rising and falling sea levels associated with climatic changes we've already discussed in the previous chapter.

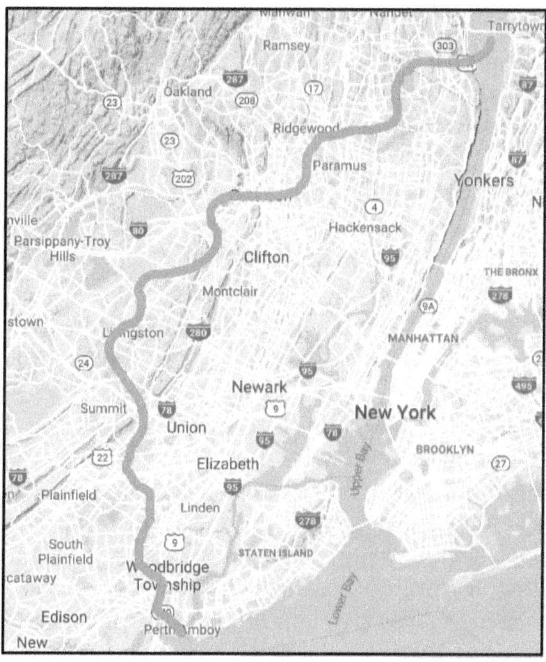

Possible course of the Hudson River through New Jersey prior to the most recent glaciation.

It's reasonable to ask how, even when following weaker fault zones, a river can cut through ridges of resistant rock like those in the Highlands, Palisades, or Watchung Mountains. The answer is due to the fact that they weren't resistant ridges back then but buried rock strata. The drainage systems which formed between the time of the breakup of Pangaea and the Pleistocene glaciation, a period of over 150 million years, existed within an entirely different landscape from that we see today. Layers of sedimentary rock that once existed in the Hudson Valley are completely eroded away today. More recent coastal plain sediments were once much thicker and more laterally extensive than the remnants existing today. The river systems in this environment essentially flowed over top of the rocks which are now ridges, but were then buried beneath the ancient land surface. Over time, as erosion continued, the rivers cut down through these resistant rocks, changing course as needed to exploit weaker areas, until we eventually see a river neatly flowing through a narrow notch in a prominent ridge today.

These so-called water gaps, also called wind gaps if a river no longer flows through them, are common in the Valley and Ridge of Pennsylvania where streams cut down through the ancient, flatter land surface to incise what are now resistant ridges of sedimentary rock. One of the most famous examples of this is the Delaware Water Gap along Interstate 80 as it passes from New Jersey into Pennsylvania. The gap formed as the Delaware River cut a narrow notch through the Kittatiny Ridge – the southern extension of the Shawangunk Ridge in New York.

One important concept in fluvial geomorphology (the study of how rivers shape the landscape) is that of base level. If a river flows into the sea, it can only downcut to sea level. It can't cut below that base level since it would then have to flow uphill to get into the ocean (another rather obvious concept in fluvial geomorphology is that water only flows downhill). When a river is high above base level, due to uplift of the land surface or dropping of sea level, it has more gravitational potential energy and will erode and downcut more rapidly. If a river is close to base level, due to erosion of the land surface or a rising of sea level, it has less energy and will tend to form low-gradient, meandering floodplains. If sea level rises enough, it may even form a drowned river valley with seawater inundating the former river system. Looking at a satellite image of Chesapeake Bay, for example, will clearly reveal its origin as a dendritic (branching) system of river valleys which were drowned by rising sea levels after melting of the Wisconsin stage glaciers.

The Hudson today is also a drowned river. The lower reaches of the river are brackish, a mixture of fresh and salt water, and daily tidal fluctuations extend to north of Albany. Drowned river valleys are called estuaries and the Hudson is even sometimes referred to as a fjord, a Norwegian word denoting a drowned glacial valley – especially where it flows through the steep-walled gorge of the Hudson Highlands. Estuaries are typically rich ecosystems, with a mixing of freshwater and saltwater species, and the Hudson is no exception. During the early years of European settlement, whales were sometimes spotted in the river and, even today, harbor seals (*Phoca vitulina*) and bottlenose dolphins (*Tursiops truncatus*) can occasionally be spotted in the river. I was recently surprised to learn that seahorses (*Hippocampus erectus*) can even be found in the harbor. All told, over 200 species of fish are represented in the Hudson watershed.

During the Pleistocene Epoch, with its multiple stages of glaciations and interglacials, sea levels fluctuated up and down significantly over time scales of thousands of years. When sea levels were low, the Hudson carved a river valley far out onto the continental shelf. At the edge of the shelf, at the continental slope, this valley becomes an impressive canyon running down into the deep abyssal plains. While the shallower valley was carved during the height of the Wisconsin glaciation, when sea levels were over 100 meters (300 feet) lower than they are today and the shoreline was tens of kilometers further out on the continental shelf, the deeper underwater canyon would have been primarily carved by the massive volumes of water brought down the Hudson Valley after the melting of the glaciers. Especially potent would have been the walls of water pouring down the valley when large post-glacial lakes, like Lakes Iroquois and Albany, broke through their dams and relatively quickly drained into the Atlantic.

It's virtually impossible to talk about the human history of the Hudson Valley without acknowledging its geologic history. The river and its tributaries have hosted human settlements in their valleys since the earliest Native Americans migrated in following the retreat of the glaciers. For much of human history in this area, these rivers also formed "highways" for travel and transport of goods throughout the region. Native Americans and the early Europeans naturally settled in the broad, fertile river valleys. Geology thus indirectly influenced human settlement and migration throughout the area.

The geography of the river also played a vital role in the Revolutionary War. As mentioned in previous chapters, the location of fortifications in the Hudson Highlands was primarily due to their strategic importance given the narrowing of the river in this area. If you look at a map of the Hudson River at West Point, for example, you'll see that the river makes almost a right-angle turn here between the Point and Constitution Island on the east. Ships sailing up the river would have to slow down here making them vulnerable to attack. This was also one of the two locations where an iron chain was stretched across the river (the other location was at Fort Montgomery near the modern Bear Mountain Bridge). These chains, as well as cannons and other armaments of war, were locally forged from Hudson Highlands magnetite ore. Further north, the narrowing of land on the western side of the river near the Stark's Knob outcropping of Ordovician pillow basalts played a decisive role in the Battles of Saratoga as was also discussed earlier.

The 19th century saw explosive growth in New York City with the population rising, mainly through immigration, from around 60,000 people in 1800 to 3.5 million in 1900. The geology of the Hudson Valley played a role in this historical development as well by providing the resources to develop the infrastructure required for an explosive population growth of over 5,000%. Triassic brownstones from the

Newark Rift, Devonian bluestones from the Catskills, Silurian natural cement from Rosendale, and bricks from Pleistocene lake clays all contributed to the building of New York City. The river itself allowed these materials to be quickly and cheaply shipped from their source to where they were needed.

In addition, the construction of the Delaware and Hudson Canal in the early 1800s allowed cheap anthracite coal to be economically shipped from the coalfields of Pennsylvania northeastward to the Hudson and thence to New York City (and beyond). This coal fueled the engines of the Industrial Revolution and helped turn New York into an economic powerhouse. For good or ill, the Hudson Valley wasn't blessed with coal deposits since rocks of the Carboniferous Period never formed here, but the geography of the wider region allowed us to exploit those resources in Pennsylvania.

The post-glacial landscape of the region also played an important role in the development of the Hudson Valley. Mountain streams in the Catskills, Shawangunks, and Highlands were ideal sites for water-powered mills. Shawangunk conglomerate provided millstones for grinding grains and local forests provided timber for the sawmills. Starting in the early 1900s, some of these mountain streams in the Catskills were dammed to form reservoirs of fresh water which was then piped to New York City through hundreds of miles of subsurface aqueducts. The higher elevation of these reservoirs allows gravity to do all the work of transporting the water southward – even as the Catskill Aqueduct travels over 300 meters (over 1,000 feet) beneath the Hudson River at Storm King Mountain. According to the Department of Environmental Protection (DEP), New York's reservoir system supplies over 280 million liters (over 1 billion gallons) of water each day to the city! New York City doesn't pull its drinking water out of the Hudson because the river, as mentioned earlier, is a tidal estuary with brackish water in its lower reaches.

If you're wondering how far up the river the salt water extends, there's no simple answer to that question. There is actually a gradient of salt (NaCl) concentration from the saline waters of the Atlantic up to the fresh waters of the river. Generally, the upper reach of the saline waters is defined by the salt front – the location along the river where the concentration is 100 milligrams of chloride ions per liter of river water (100 mg/L of Cl-). The salt front is constantly in motion. In spring, snowmelt and rains move it to the south, generally near the Tappan Zee Bridge, and it tends to reach its maximum northward position near Newburgh in late summer. The position also obviously depends upon whether it's a particularly wet or dry year. Large precipitation events can quickly move the salt front downriver and occasional droughts bring it further northward.

This is why New York City's emergency drinking water supply from the Hudson is located at the Chelsea pump station north of Newburgh (on the east side of the river). This was a convenient location because it's here that the Delaware Aqueduct crosses beneath the Hudson River. During a severe drought, however, the salt front may well extend even beyond this pump station possibly requiring some dilution of the river water with other sources of fresh water to reduce its salinity to acceptable levels for drinking.

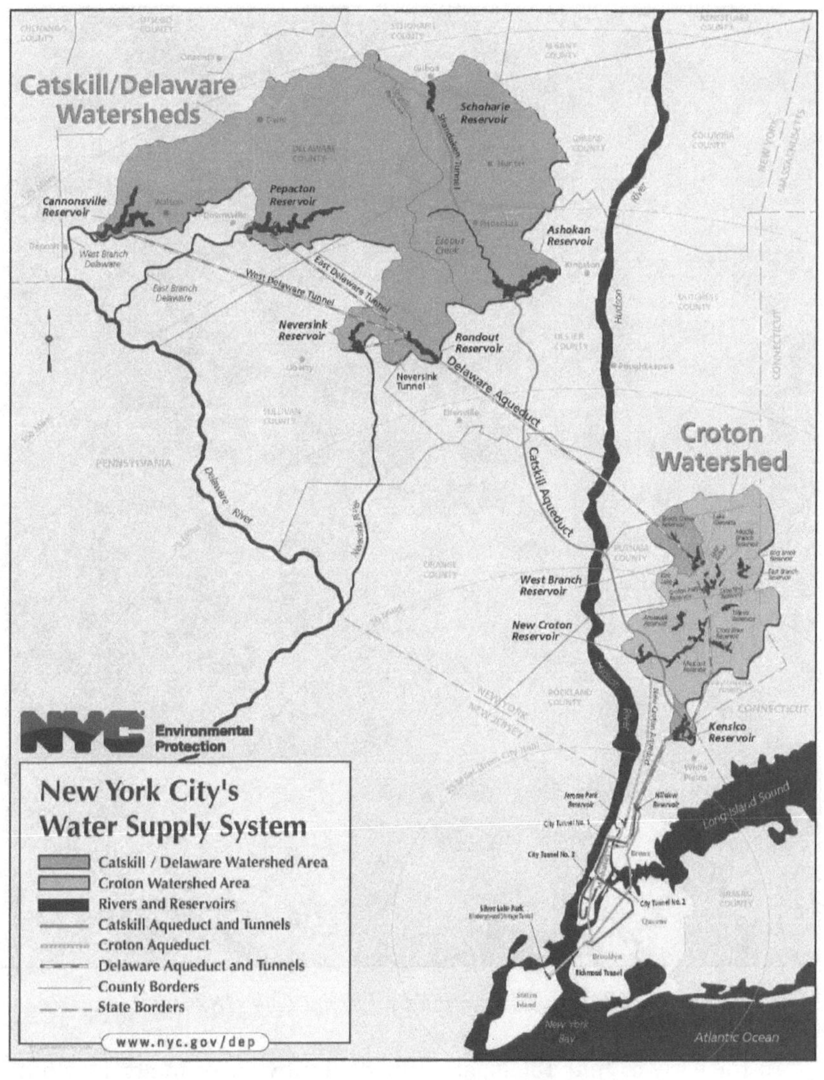

New York City reservoir and aqueduct system

The natural beauty of the Hudson River Valley is a resource that's a bit harder to quantify, but just as valuable in today's world as were the economic resources of the past. The Hudson Valley has long been appreciated for its scenic qualities. On the 1609 voyage of Henry Hudson up the river, ship's officer Robert Juet described it in an understated way as "…a pleasant land to see." One can only imagine what the view from the river must have looked like then without all the present-day development lining the shores (although there were plenty of Native American settlements at the time).

Falls of the Kaaterskill by Thomas Cole (1826)

In the early to mid-1800s, an influential movement in art developed as an outgrowth of Romanticism in art, literature, and music. It's now called the Hudson River School, although it refers to a style of painting rather than any organized institution. While many of the Hudson River School paintings showed scenes in the Hudson Valley, the artists associated with this movement also painted in such far-flung places as Niagara Falls, the Adirondacks, the White Mountains, the American West, Europe, and even the Andes of South America. The art of the Hudson River School concentrated on realistic, although not always completely accurate, representations of natural landscapes on a grand scale. Humans were generally shown in such paintings as small and insignificant figures dwarfed by the overwhelming scenery.

The Hudson River School is usually acknowledged to have begun around 1825 with Thomas Cole (1801-1848), born in England but spending much of his adult life at his studio in Catskill, New York. According to Cole, "The subject of art should be pure and lofty ... a moral, religious or poetic effect must be produced on the mind." This was a time when nature was starting to be appreciated for its own sake. This may, in part, have been a revolt against the excesses and environmental degradation of the Industrial Revolution but this economic growth also provided enough wealth for people to travel and experience these areas of scenic beauty. Other Hudson River School painters include Asher Durand (1796-1886), a friend of Thomas Cole, Frederic Church (1826-1900), a student of Cole's and owner of the Persian-style estate of Olana across the river from Catskill, and Jasper Cropsey (1823-1900), best-known for the use of vibrant colors in his Hudson Valley autumn landscapes.

Tourism developed in the Hudson Valley during the 19th century, partly due to the advertising of the area's beauty by the Hudson River School paintings and the increasing ease of traveling upriver in steamboats from New York City. Resort hotels, like the Catskill Mountain House, exposed many people to the region and an appreciation of the natural landscape eventually led to the preservation of many areas within the Valley. Today, there are dozens of State and local parks located throughout the area and outdoor recreation and tourism is an important component of the local economy.

Increased population growth and urban sprawl in the area is, of course, placing pressure on our remaining resources. Fortunately, many people are paying attention to such issues as overdevelopment, habitat loss, and even how various proposed projects might affect the "viewshed" of the valley. No one wants to come to a beautiful overlook in the mountains surrounding the Hudson only to overlook an industrial park. I think most residents of the Hudson Valley would acknowledge the importance of open space and that the scenic beauty of the area is important to preserve.

An inevitable result of human development is, unfortunately, environmental degradation and many of the environmental issues in the Hudson Valley do have a geological component. There's a mine, for example, located a short distance from the Appalachian Trail in the Hudson Highlands and overlooking the Bear Mountain Bridge where several acres of barren land exist within the middle of a forest. This mine exploited both iron- and copper-bearing minerals but has not been operational in any significant way for at least 100 years. The problem arose because one of the dominant waste minerals left behind in large waste piles was pyrrhotite, an iron sulfide mineral. Sulfides (metals bonded to sulfur atoms) chemically react with water and oxygen to form sulfuric acid (H_2SO_4) and this led to highly acidic groundwater draining from the mine area – a problem called, appropriately enough, acid mine drainage. Decades later, the only plants growing in the immediate area are a few small, stunted birch trees which can tolerate acidic soils. This is similar to what can be seen in some coal mining areas in Pennsylvania with similar acid mine drainage problems.

Another problem with acidity arises from air pollution rather than groundwater contamination. Burning fossil fuels generates sulfur and nitrogen oxides which then chemically interact with atmospheric water to form sulfuric (H_2SO_4) and nitric (HNO_3) acids. This process leads to acid rain (actually acid precipitation since it affects snow as well) which may fall hundreds of kilometers away from the original source of the air pollution. While many people are aware of acid rain problems in the Adirondacks, it also affects the entire Hudson Valley region. In the previous chapter, the five sky lakes of the Shawangunks (Mohonk, Minnewaska, Awosting, Haseco, and Maratanza) were discussed in the context of glacial erosion. Acid rain over the years has effectively killed off almost all plant and animal life in four of these five lakes. While the lakes are beautiful jewels of the Shawangunks, they're also biologically dead.

The geologic factor all of these lakes have in common is that they're within the Shawangunk Conglomerate. This formation is almost pure quartz, an unreactive mineral, and once lake water acidifies in such a setting, it will essentially stay acidic (it's like placing the water in a glass beaker). The only exception is Lake Mohonk where the conglomerate was somewhat thinner and the bottom of the lake is underlain by the Martinsburg Group shales. Shale will chemically buffer the water making it less acidic and thus able to support fish and other aquatic life that's lacking in the other lakes. Once again geological conditions, in this case bedrock lithology, affect an environmental condition.

Groundwater contamination is a significant problem in some areas of the Hudson Valley as well. The small town of High Falls located along the banks of the Rondout Creek in Ulster County, for example, is an Environmental Protection Agency (EPA) Superfund site. A series of now defunct factory operations was located just outside of town and utilized a number of volatile organic compounds (VOCs) as solvents, primarily for degreasing and painting, starting in the early 1960s. These solvents, through a combination of poor practices and the lax environmental laws of the time, leached into the soil behind the factory. In time, these contaminants moved downward through fractures in the bedrock – the Shawangunk Conglomerate – and became part of the local groundwater system. Groundwater, along with the plume of contamination, slowly flowed toward the Rondout Creek through the interconnected fractures within the bedrock. This is why a knowledge of geology is essential to understanding such issues since geologists can best predict the regional orientation of such subsurface fractures.

The contaminants inevitably encountered drilled wells and were pulled up into people's homes where they were ingested over the course of many years. In 1994, a homeowner in the area noticed a "chemical" smell to their water when the concentration of the contaminants reached high enough levels. Testing revealed the presence of chemicals with intimidating names like 1,1,1-trichloroethane, trichloroethylene, 1,1-dichlorethylene, and 1,1-dichloroethane in the water. All of these chemicals are obviously harmful to humans and this led to the involvement of the EPA. More than a decade after first being discovered, and after spending millions of dollars of taxpayer's money, the residents of High Falls are now on a municipal water system. Their water is now taken from the Catskill Aqueduct and their backyard wells are capped.

The EPA is currently pumping and treating the polluted water from the aquifer under High Falls in an attempt to keep the plume of contamination from entering the nearby Rondout Creek. Treatment is done by a process called air stripping where the water is sprayed in a fine mist and the VOC vapors are trapped in carbon filters. The treated, and presumably clean, water is then released into a nearby stream. Unfortunately, chemicals in the groundwater system don't behave like chemicals moving through a simple pipe. In a pipe, we could simply run some water through and flush out the contaminant. In the groundwater system, the contaminating chemicals often interact with clay minerals lining fractures within the rock and remain in the aquifer for very long periods of time with constant, gradual leakage into the groundwater. The pumping and treatment of groundwater in High Falls will have to be done for decades. Unfortunately, this story of aquifer contamination can be repeated over and over again, not only at other locations in the Hudson Valley, but all around our polluted world.

The Hudson River itself has been greatly affected by pollutants since the Industrial Revolution. By the middle of the 20th century, the river was unsafe to swim in, let alone drink, and was often described as an 'open sewer.' This wasn't too much of an exaggeration. Billions of gallons a day of raw sewage were being dumped into the river by many municipalities lining its shores. New York City was the biggest offender. The problem is that most cities have combined storm water and sewage drainage systems. The sewage treatment plants in those municipalities simply can't cope with the increased volume of water during storm events and end up routing raw sewage directly into the Hudson. There's no easy solution to this problem.

Industrial wastes commonly found their way into the river as well. For many years, the mantra was "dilution is the solution to pollution." In other words, pollutants dumped into a river are diluted by the massive volume of flowing water and harmlessly carried away. This may work on a small scale, but simply not true in the 20th century when factories lined the banks of the Hudson from New York Harbor northward to the Adirondacks. Pollution became noticeable to even casual observers of the Hudson. Many people still remember that a few decades ago, for example, it possible to tell what color the General Motors plant in Tarrytown was using to paint their vehicles that day by the hue of the river downstream of their facility.

Near Foundry Cove in Cold Spring, just north of Constitution Marsh, the Marathon Battery Company (this plant went through a number of company name changes) manufactured batteries both for the commercial market as well as having large contracts with the United States military starting in 1952. The company discharged large amounts of nickel and cadmium hydroxides, used in the manufacturing process, not only into the Cold Springs sewer system but also directly into the waters of the cove. By 1970, Foundry Cove was the most cadmium polluted site in the world. The problem with these metals is that they don't simply dilute in the water column and wash down the river. These metals settle into the bottom sediments where they remain, essentially forever, since they don't naturally break down over time. These metals are also toxic to bottom-dwelling aquatic organisms that burrow into these sediments. After the involvement of the EPA, a number of studies, and the dredging of hundreds of thousands of tons of sediment from the area, the site was essentially remediated by the mid 1990s and much of the area is now protected.

The most infamous example of Hudson River pollution was the contamination of bottom sediments from General Electric plants in Hudson Falls and Fort Edwards. These plants manufactured large capacitors for the utility and released almost 600,000 kilograms (over 1,000,000 pounds) of polychlorinated biphenyls (PCBs) into the river from the 1950s to the 1970s. Throughout this time, much of the contamination was contained within bottom sediments trapped behind a dam at Fort Edward. When the aging dam was removed in 1973, the PCBs were then able to migrate downstream resulting in contaminated "hot spots" along the river bed. PCBs can now be detected in the entire river all the way down to New York Harbor and are of concern because they bioaccumulate in fish and humans eating those fish.

The EPA, the State of New York, environmentalist groups, and General Electric are still arguing over the best way to remediate the PCB problem decades later. Contamination of bottom sediments by heavy metals or chemical toxins is one of the most vexing environmental issues in the Hudson River. Is it better to dredge the sediment up from the river bed and dispose of them on land, and risk remobilizing the PCBs into the river water, or should we just leave them alone to naturally break down as they're buried under the sediments? While raw sewage is unpleasant, water flushes it through the river system allowing it to be cleaned up relatively quickly (although it may still leave long-lasting biological effects). Many metals and chemicals, however, can bind with clays in the sediments of the river and remain in the system for very long periods of time by human standards. Tides, storms, and biological activity can reactivate these materials into the water column even after they've been buried.

In many ways, the modern environmental movement starting in the 1960s had its roots here in the Hudson Valley. In 1962, the Consolidated Edison utility company planned to construct a massive pump storage reservoir system in the face of Storm King Mountain right on the Hudson River. To meet New York City's ever-increasing demand for electricity, Con Edison planned to pump water from the river at night when there was a surplus of electricity, up to a storage reservoir near Storm King Mountain (in the present-day Black Rock Forest Preserve). During the daytime hours, when the demand for electricity was high, the utility would release the water back into the river and generate hydroelectric power. While it's a somewhat inefficient process, a cost-benefit analysis by the utility showed that it was economically feasible at the time.

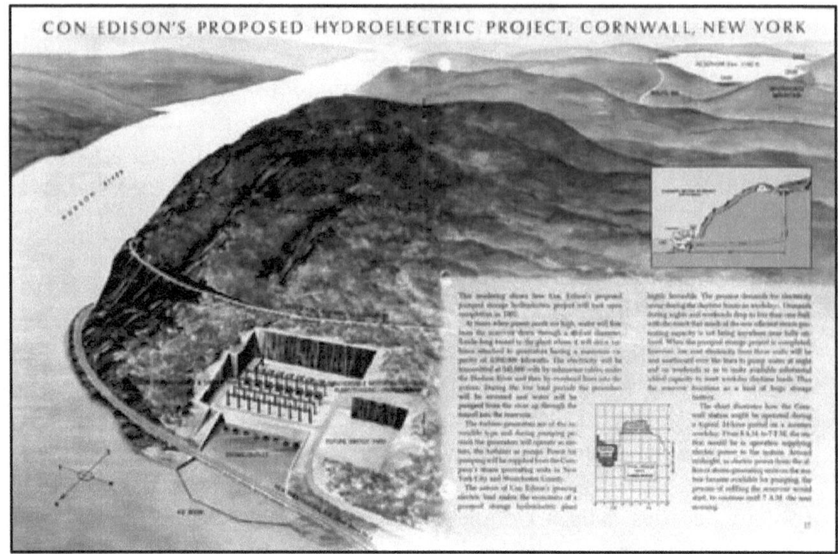

Con Edison's controversial project

Shortly after Con Edison announced their plans, a small group of private citizens, meeting in local homes, formed the Scenic Hudson group to oppose this project. Their opposition was based on a number of factors including its potential damage to Hudson River fisheries (fish don't do well when pulled into turbines) and the importance of preserving the scenic beauty and historical significance of Storm King Mountain – the northern gateway to the Hudson Highlands. It took almost 18 years of court cases, but Con Edison eventually reached a settlement agreement with Scenic Hudson, which also included some compromises regarding the Indian Point nuclear power plant. Con Edison abandoned their plans for the Storm King project.

Around this same time, in response to the increasing pollution of the Hudson, folk singer and environmental activist Pete Seeger started raising money and support for the idea of building a wooden sloop, the Clearwater, to sail the Hudson and educate people about the environment. The Clean Water Act, passed in 1972, along with lawsuits from environmental organizations like Clearwater, Riverkeeper, Scenic Hudson, and others has gradually forced the cleanup of the Hudson. While it's always depressing to discuss environmental issues, the good news is that the Hudson is, in fact, cleaner today than it's been for many years thanks to the hard work of untold numbers of concerned citizens.

We will never recover the pristine waters that the Lenape and early Dutch settlers knew, but we can all work toward ensuring that our families can feel comfortable visiting a Hudson River beach for a day of swimming or someday catch a striped bass that's safe to eat. Steve Stanne, author of *The Hudson: An Illustrated Guide to a Living River* and an environmental educator, once told a group of us out on the river that we can argue all day about where the Hudson River actually "begins" but it really has its source in all of the headwaters of all of the tributaries that flow into it along its entire length from the Adirondacks to the sea. A cigarette butt tossed into a storm drain in Newburgh, an oil leak from a car parked in a driveway in Nyack, and a soda bottle tossed into a roadside ditch in Albany all contribute to the pollution of the Hudson River. Bringing it back to the quotation opening this chapter by John Muir, and as all environmentalists know, everything in this world is connected.

Postlude
Treasures Under Foot

The lesson which life repeats and constantly enforces is "Look under foot." You are always nearer to the divine and the true sources of your power than you think. The lure of the distant and the difficult is deceptive. The great opportunity is where you are. Do not despise your own place and hour. Every place is under the stars, every place is the center of the world.

John Burroughs
The Divine Soil

John Burroughs, the great Hudson Valley naturalist, had it right. We often grow blind to the treasures right under our feet. Many of us who live in the Hudson Valley long to travel to distant, exotic locales while all too often ignoring the sublime beauty in our own backyards. John Burroughs found beauty under a piece of bark lying in the woods; I myself find it in what most people would consider to be nondescript rocks on the side of the road.

John Burroughs also valued the findings of modern science unlike some of the earlier 19th century romantics who believed that scientific inquiry and the proper appreciation of nature were somehow mutually exclusive. Suspicion and even fear of science has been around for a long time. In our modern world, animosity toward science is often rooted in political ideologies (climate change issues), religious beliefs (young-Earth creationism), or an attraction to some particular branch of pseudoscience (astrology, for example). At a more sophisticated level, some postmodernist philosophers have even criticized science as a Western social construct with no objective reality. Many times, in my experience, this opposition to science is typically rooted in ignorance of what they're actually criticizing. Unfortunately, such opposition to science is difficult to counter since these are beliefs in which people often have an emotional investment and are therefore generally not amenable to critical discussions.

What about the 19th century Romantic Movement criticism of science, one that many still believe today, that science "demystifies" the world and in some sense robs it of its beauty? Take an introductory botany class, for example, and you'll be introduced to a lot of technical terms – calyx, gynoecium, meristem, stigma, and corolla to name but a few. Does not the science of botany in this way reduce a beautiful flower into a mass of complex jargon? I would contend that this view is naïve. I think most botanists would likely tell you that their scientific knowledge of flowers increases their appreciation for them since there's also beauty to be found in understanding the evolutionary development of flowering plants over time, the complex chemical reactions that drive all living things, the symbiotic relationships that develop between flowering plants and animals, and much more. But the grounding in terminology is necessary because you can't learn about the wonders of pollination without knowing the difference between a pistil and a stamen.

Terminology is also an important part of a descriptive science like geology since it allows us to unambiguously refer to something – every geologist knows what a "limestone" is – but geology is much more than identifying rocks (just as botany is much more than labeling flower parts). It's not enough to just apply the label of limestone to a rock outcrop, but to ask how and when that limestone formed. What does the limestone teach us about the geologic history of our area? Does the outcrop contain fossils? What is the geochemical composition and microscopic structure of the limestone? What types of rocks exist above and below the limestone beds? How regionally extensive is this limestone? Is the limestone deformed by later folding or faulting? After answering each of these questions, it's not simply a limestone anymore. It's a regionally-extensive, Early Devonian Period, fossiliferous, marine limestone representing deposition in the subtidal environment of a warm, shallow sea and which was later folded by the Acadian mountain building event. We now have a context for that limestone and it's teaching us about a slice of ancient time in our area.

It's also interesting to relate that limestone outcrop to a larger picture. How does the limestone bedrock affect the type of flora and fauna found in the area? How do any solution features in the limestone affect the local groundwater system? What type of human activity is present in the area and how might that affect the limestone aquifer? Has the limestone been exploited for crushed rock or cement quarrying and therefore influenced the economic development and history of the area? These are all important questions and can't even be asked, let alone answered, without an understanding of the terminology and basic concepts of geology. It's similar to the way one needs to master the rules of arithmetic before they can appreciate the deeper beauty of number theory. Far from limiting, the terminology allows you to extend the questions that you can ask.

This need to introduce terminology in descriptive sciences does cause problems. As a community college professor, I come across many students who tell me they "hate" science because it's boring or lacks any relevancy to their lives. To them, science is simply memorizing a long list of difficult-to-pronounce terms and disconnected concepts. While this may be what they need to pass a standardized exam in most public school systems, it's not really learning science. Science is a method of investigating and understanding the world around us and the best way to learn science is to actually do science. In the Earth sciences, that means getting students out to examine real rock outcrops. Unfortunately, some administrators, who themselves often lack a scientific background, fail to understand this concept and many of my geology colleagues around the country are constantly battling for the resources necessary to get students out into the field.

Most of the questions I posed about the hypothetical limestone simply can't be answered by handing a student a small piece of limestone in a geology classroom. They can only be answered in the field while crawling around on that limestone outcrop. To go back to the Prelude to this book, rocks are best studied by pressing your nose up against the outcrop. Unfortunately, until students take a geology course in college, they typically don't ever have the opportunity to go on a field trip to examine the local rocks. In most public schools, Earth science is taught entirely indoors and, many times, with no relevance to student's lives. So while there's a need to understand the basic terminology and concepts, it also needs to be related to the real world and that's where traditional science teaching often fails.

It's also funny how something that appears completely irrelevant to a student can become very relevant years later as an adult. It may be boring to talk about soil porosity and permeability in a geology classroom but it suddenly becomes interesting when several thousand dollars are added to the cost of your new house because the soil in your backyard failed a percolation test and you need a specially-engineered septic system. The concept of flood plains becomes real when you can't get affordable flood insurance because your house is located in a 50-year flood zone (or worse yet, actually gets flooded as many have here in Ulster County). Some knowledge of geology would be useful to people currently building expensive homes on geologically-recent volcanic mudflows near Mount Rainier in Washington State (a volcano which has a 100% chance of erupting again).

And not everything needs to have a practical application. Sometimes the relevance is simply that the knowledge awakens wonder in our hearts. Scientific knowledge enhances our awe of the natural world. Without the science of geology, they are just gray rocks on the side of the road. Big deal. Maybe some sparkle in the sunlight or are shaped like an animal but such observations are trivial when we can look at these same rocks and see evidence for an ancient sea teeming with life in the dim mists of time, Himalayan-scale mountains weathering away grain-by-grain for tens of millions of years, and continents drifting over the entire surface of the globe constantly rearranging the Earth's landscapes. There may not be much practical benefit in knowing this but we'd be much poorer in our ignorance.

If you've read to this point, I hope you've gained an increased appreciation for this beautiful area we call home and its long and interesting geologic history. The story I've told here is only an outline and each chapter contains many topics that could easily be expanded upon in far more detail. I urge you to further explore those areas that interest you. Above all, get out and look under foot with open eyes. Wherever your path leads, I wish you an enjoyable journey.

Appendix A
Selected Bibliography

Books to the ceiling, books to the sky.
My piles of books are a mile high.
How I love them! How I need them!
I'll have a long beard by the time I read them.

Arnold Lobel
Whiskers & Rhymes

Where did all of the information in this book come from? It's a complex mélange of having grown up in the area; being a geology student at three Hudson Valley colleges and universities – Ulster County Community College, SUNY New Paltz, and the University at Albany; the reading of countless geology books and professional papers; discussion with colleagues at conferences and on the outcrop; and innumerable field trips with students, faculty, and knowledgeable local residents over the past few decades.

I can, however, refer the interested reader to a number of books and field guides with additional information on the regional geology. I'll also mention that I regularly teach a summer field course at SUNY Ulster County Community College called *Geology of the Hudson Valley* that's open to anyone willing to participate, able to hike around some beautiful areas, and not afraid to get a little dirty.

The Hudson Valley is also blessed with many incredible nature preserves and state parks, most of which have educational programs and materials to learn more about their natural history.

Ansley, Jane Elizabeth. 2000. *The Teacher-Friendly Guide to the Geology of the Northeastern U.S.* The Paleontological Research Institute: Ithaca, NY.

Dunwell, Frances F. 2008. *The Hudson: America's River.* Columbia University Press: New York, NY.

Evers, Alf. 1984. *The Catskills: From Wilderness to Woodstock: Revised and Updated.* Overlook Press: Woodstock, NY.

Evers, Alf, Titus, Robert, & Weidner, Tim. 2008. *Catskill Mountain Bluestone.* Purple Mountain Press: Fleischmanns, NY.

Fagan, Jack. 2006. *Scenes and Walks in the Northern Shawangunks* (3rd edition). New York–New Jersey Trail Conference: Mahwah, NJ

Fisher, Donald W. 2006. *The Rise and Fall of the Taconic Mountains: A Geological History of Eastern New York.* Black Dome Press: Hensonville, NY.

Hutton, George V. 2003. *The Great Hudson River Brick Industry: Commemorating Three and a Half Centuries of Brickmaking.* Purple Mountain Press: Fleischmanss, NY.

Isachsen, Y. W., Landing, E., Lauber, J. M., Rickard, L. V., and Rogers, W. B. (editors). 2000. *Geology of New York: A Simplified Account* (2nd edition). New York State Museum Educational Leaflet 28: Albany, NY.

Lenik, Edward J. 1996. *Iron Mine Trails: A History and Hiker's Guide to the Historic Iron Mines of the New Jersey and New York Highlands.* New York – New Jersey Trail Conference: Mahwah, NJ

Millhouse, Barbara Babcock. 2007. *American Wilderness: The Story of the Hudson River School of Painting.* Black Dome Press: Hensonville, NY.

Raymo, Chet & Raymo, Maureen E. 2008. *Written in Stone: A Geological History of the Northeastern United States* (3rd edition). Black Dome Press: Hensonville, NY.

Rogers, William B., Isachsen, Yngvar W., Mock, Timothy D., and Nyahay, Richard E. 1990. *New York State Geological Highway Map.* New York State Museum Educational Leaflet 33: Albany, NY

Titus, Robert. 1993. *The Catskills: A Geological Guide.* Purple Mountain Press: Fleischmanss, NY.

Titus, Robert. 1996. *The Catskills in the Ice Age.* Purple Mountain Press: Fleischmanss, NY.

Van Diver, Bradford B. 1985. *Roadside Geology of New York.* Mountain Press Publishing: Missoula, MT.

Van Zandt, Roland. 1997. *The Catskill Mountain House.* Black Dome Press: Hensonville, NY.

New York State Geological Association (NYSGA) Field Guides.
Geological Society of America (GSA) Abstracts with Programs.

Illustration Credits

Prelude
2   Brachiopod and bivalve shells drawing by Jennifer Wulfe
3   *Gypidula* drawing by Jennifer Wulfe
4   Crinoid drawings by Jennifer Wulfe
5   *Favosites* drawing by Jennifer Wulfe

Chapter 1
14   Periodic table illustration by Steven Schimmrich
15   Radioactive decay illustration by Steven Schimmrich
19   Stromatolite drawing by Jennifer Wulfe

Chapter 2
25   Crust and mantle illustration by Steven Schimmrich
26   Earth's tectonic plates illustration by Steven Schimmrich and Jennifer Wulfe
28   Three tectonic plate boundaries illustration by Steven Schimmrich
29   Continental volcanic arc illustration by Steven Schimmrich and Jennifer Wulfe
30   India approaching Asia illustration by Steven Schimmrich and Jennifer Wulfe
32   Rodinia illustration by Steven Schimmrich and Allison Schimmrich. Modified from the reconstruction of Hoffman, 1991, *Science* 252: 1409-1412
33   Extent of Grenville-age rocks illustration by Steven Schimmrich and Jennifer Wulfe. Modified from Volkert, et al., 2010, *Memoir of the Geological Society of America* 206:307-346
36   Gneissic layering and hand lens drawing by Jennifer Wulfe
38   Fault and shear zone illustration by Steven Schimmrich

Chapter 3
44   Nicolas Steno and shark teeth illustrations are public domain at Wikipedia [https://en.wikipedia.org/wiki/Nicolas_Steno]
46   Cross section of a river valley illustration by Steven Schimmrich and Jennifer Wulfe
48   Geologic time scale illustration by Steven Schimmrich. Age dates from Walker, et al., 2018 *Geologic Time Scale v. 5.0*, Geological Society of America.
49   Continental rifting illustration by Steven Schimmrich and Jennifer Wulfe
50   Mid-ocean ridge illustration by Steven Schimmrich and Jennifer Wulfe
53   Laurentia at 510 Ma illustration by Steven Schimmrich and Jennifer Wulfe. Traced and modified from the paleogeographic reconstructions of Ron Blakey, 2013, Colorado Plateau Geosystems, Inc.
58   Wappinger Group outcrop photo by Steven Schimmrich
59   Trilobite drawing by Jennifer Wulfe
60   Lester Park stromatolites photo by Steven Schimmrich

Chapter 4
63   West coast tectonics illustration by Steven Schimmrich and Jennifer Wulfe
64   Disconformity illustration by Steven Schimmrich and Jennifer Wulfe
68   Laurentia at 470 Ma illustration by Steven Schimmrich and Jennifer Wulfe. Traced and modified from the paleogeographic reconstructions of Ron Blakey, 2013, Colorado Plateau Geosystems, Inc.
69   Aleutian island arc illustration by Steven Schimmrich and Jennifer Wulfe
70   Clay minerals drawing by Jennifer Wulfe
71   Graptolites drawing by Jennifer Wulfe

191   Pangea Politica illustration public domain by designer Massimo Pietrobon [https://capitan-mas-ideas.blogspot.com/2012 08/ pangea-politica.html]

192   Glossopteris leaf fossil (cropped) is public domain at Wikipedia [https://en.wikipedia.org/wiki/Glossopteris]

193   Mesosaurus fossil photo (cropped) by "Tommy from Arad" at Wikipedia – CC BY 2.0 [https://en.wikipedia.org/wiki/Mesosaurus]

194   Angle of inclination illustration by Steven Schimmrich

195   Satellite image of the Valley and Ridge from Google Earth

197   Physiographic provinces of New York illustration modified from Mulvihill & Baldigo, 2012, *JAWRA*, 1-15.

199   Neuropteris photo (cropped and rotated) by Gunnar Ries at Wikipedia – CC BY-SA 2.5 [https://en.wikipedia.org/wiki/ Neuropteris]

200   Lepidodendron reconstruction by "Falconaumanni" at Wikipedia. – CC BY-SA 3.0 [https://en.wikipedia.org/wiki/Lycophyte]

200   Club moss photo by Steven Schimmrich

Chapter 11

207   East African Rift System illustration by the USGS is public domain [https://pubs.usgs.gov/gip/dynamic/East_Africa.html]

208   Normal and reverse fault illustrations by Steven Schimmrich

211   Rifting of Pangaea illustration by the USGS is public domain [https://www.usgs.gov/media/images/nyc-region-breakup-pangaea-formation-mesozoic-rift-basins]

213   Brownstones photo is public domain at Wikipedia [https://en.wikipedia.org/wiki/Brownstone]

215   Hammer Creek Conglomerate photo by Steven Schimmrich

216   Palisades Sill photo by Steven Schimmrich

219   *Coelophysis bauri* (modified) illustration by Jeff Martz/NPS is public domain at Wikipedia [https://en.wikipedia.org/wiki/Coelophysis]

220   Theropod track photo by Allison Schimmrich

Chapter 12

223   Worldwide mid-ocean ridge illustration by J. M. Watson / USGS is public domain at Wikipedia [https://en.wikipedia.org/wiki/Mid-ocean_ridge]

224   Structure of the oceanic crust illustration by Steven Schimmrich

228   Reptile versus mammalian teeth drawing by Jennifer Wulfe

230   Luis and Walter Alvarez photo by U.S. Government is public domain at Wikipedia [https://en.wikipedia.org/wiki/ Luis_Walter_Alvarez]

231   Chicxulub illustration by Steven Schimmrich

Chapter 13

237   Foraminifera photo by "Wilson44691" is public domain at Wikipedia [https://en.wikipedia.org/wiki/Quinqueloculina]

243   Chatter marks photo by Steven Schimmrich

245   Umpire's Rock photo by Steven Schimmrich

246   Patterson's Pellet photo by Steven Schimmrich

250   Draining of Lake Iriquois illustration by Jack Cook, 2005, *Oceanus Magazine*, Woods Hole Oceanographic Institution

251   Clay pit photo credit unknown. Assumed public domain.

252   Mammoth and mastodon teeth drawing by Jennifer Wulfe

253   Cohoes mastodon photo by Allison Schimmrich

254   Hyde Park mastodon photo by Allison Schimmrich

About the Author

Steven Schimmrich resides in the mid-Hudson Valley and is a Professor of Earth Sciences and Geology at SUNY Ulster County Community College in Stone Ridge, New York where he's been employed since 1999. In 2007, he won the State University of New York Chancellor's Award for Excellence in Teaching.

He can often be found wandering the wild places of the Hudson Valley.

About the Illustrator

Jennifer Wulfe is a Hudson Valley artist. Her work ranges from scientific illustration to whimsical art using pencil, watercolor, ink, and acrylics.

She is very excited to see the culmination of this book for its filling the void in this particular niche of Hudson Valley history, and as an academic and personal achievement for Steven.

Steven H. Schimmrich